Rat Lungworm Disease

Rat Lungworm Disease

Guest Editor

Susan Jarvi

MDPI

Basel • Beijing • Wuhan • Barcelona • Belgrade • Novi Sad • Cluj • Manchester

Guest Editor
Susan Jarvi
Pharmaceutical Sciences
University of Hawaii Hilo
Hilo
United States

Editorial Office
MDPI AG
Grosspeteranlage 5
4052 Basel, Switzerland

This is a reprint of the Special Issue, published open access by the journal *Pathogens* (ISSN 2076-0817), freely accessible at: www.mdpi.com/journal/pathogens/special_issues/Rat_Lungworm_Disease.

For citation purposes, cite each article independently as indicated on the article page online and using the guide below:

Lastname, A.A.; Lastname, B.B. Article Title. *Journal Name* **Year**, *Volume Number*, Page Range.

ISBN 978-3-7258-3298-9 (Hbk)
ISBN 978-3-7258-3297-2 (PDF)
https://doi.org/10.3390/books978-3-7258-3297-2

Cover image courtesy of Jack Jeffrey Photography
Parmarion martensi (semi-slug). A prominent intermediate host of *Angiostronylus cantonensis* in Hawaii.

Contents

About the Editor

Susan Jarvi

Guest Editor Susan I. Jarvi is a Professor in the department of Pharmaceutical Sciences where she teaches Pharmaceutical Immunology and other infectious disease-related subjects at the Daniel K. Inouye College of Pharmacy at the University of Hawaii in Hilo, HI. She studies host–parasite and parasite–parasite interactions toward a better understanding of the impacts of infectious disease and the evolution of disease susceptibility. Her laboratory is ideally situated on east Hawaii Island for studies of neuroangiostrongyliasis or rat lungworm disease, as this geographic region is a national hotspot for rat lungworm disease in the USA. Her lab has conducted and continues to conduct numerous studies regarding the prevention, transmission, prevalence, diagnosis, and treatment of this pathogen and the potentially fatal disease it causes.

Preface

The pathogenic nematode *Angiostrongylus cantonensis* is a leading global causative agent of eosinophilic meningitis in humans and other species. Clinically, this disease is known as neuroangiostrongyliasis but is colloquially recognized as rat lungworm disease. This disease has resulted in morbidity and mortality in humans and other accidental hosts, and the geographic range of this pathogen continues to increase through global expansion. The first five studies in this reprint provide updates on recent global research efforts and distribution, including studies from India, Brazil, Ecuador, reports of coinfections in Spain, and the genetic-based determination of global spread patterns of *A. cantonensis*. The other six studies evaluate potential mechanisms for prevention, such as creating barriers to decrease the mollusk infestation of crops; disease diagnostics applying both DNA and antibody-based methods; and current approaches to the pharmacological management of this disease using a multi-drug approach in humans, as well as investigating the effects on *A. cantonensis* of a commercially available antiparasitic currently used for controlling heartworm, fleas, and ticks in dogs.

This reprint is derived from a previous special rat lungworm issue of the journal *Pathogens*. We showcase the valuable research work providing the most current and fascinating information available regarding the geographic distribution, prevention, diagnosis, and treatment of *A. cantonensis* infections. This collective body of the literature greatly advances our critical knowledge of *A. cantonensis* and this disease.

Susan Jarvi
Guest Editor

pathogens

Review

An Overview of *Angiostrongylus cantonensis* (Nematoda: Angiostrongylidae), an Emerging Cause of Human Angiostrongylosis on the Indian Subcontinent

Divakaran Pandian [1,*], Tomáš Najer [1] and David Modrý [1,2,3,*]

[1] Department of Veterinary Sciences, Faculty of Agrobiology, Food and Natural Resources/CINeZ, Czech University of Life Sciences Prague, 16500 Prague-Suchdol, Czech Republic; najer@af.czu.cz
[2] Department of Botany and Zoology, Faculty of Science, Masaryk University, 61137 Brno, Czech Republic
[3] Biology Center, Institute of Parasitology, Czech Academy of Sciences, 37005 Ceske Budejovice, Czech Republic
* Correspondence: pandian@af.czu.cz (D.P.); modry@af.czu.cz (D.M.); Tel.: +420-608-388-352 (D.P.); +420-724-334-808 (D.M.)

Abstract: Human angiostrongylosis is an emerging zoonosis caused by the larvae of three species of metastrongyloid nematodes of the genus *Angiostrongylus*, with *Angiostrongylus cantonensis* (Chen, 1935) being dominant across the world. Its obligatory heteroxenous life cycle includes rats as definitive hosts, mollusks as intermediate hosts, and amphibians and reptiles as paratenic hosts. In humans, the infection manifests as *Angiostrongylus* eosinophilic meningitis (AEM) or ocular form. Since there is no comprehensive study on the disease in the Indian subcontinent, our study aims at the growing incidence of angiostrongylosis in humans, alongside its clinical course and possible causes. A systematic literature search revealed 28 reports of 45 human cases from 1966 to 2022; eosinophilic meningitis accounted for 33 cases (75.5%), 12 cases were reported as ocular, 1 case was combined, and 1 case was unspecified. The presumed source of infection was reported in 5 cases only. Importantly, 22 AEM patients reported a history of eating raw monitor lizard (*Varanus* spp.) tissues in the past. As apex predators, monitor lizards accumulate high numbers of L3 responsible for acute illness in humans. For ocular cases, the source was not identified. Most cases were diagnosed based on nematode findings and clinical pathology (primarily eosinophilia in the cerebrospinal fluid). Only two cases were confirmed to be *A. cantonensis*, one by immunoblot and the other by q-PCR. Cases of angiostrongylosis have been reported in Delhi, Karnataka, Kerala, Maharashtra, Madhya Pradesh, Puducherry, Telangana, and West Bengal. With a population of more than 1.4 billion, India is one of the least studied areas for *A. cantonensis*. It is likely that many cases remain undetected/unreported. Since most cases have been reported from the state of Kerala, further research may focus on this region. Gastropods, amphibians, and reptiles are commonly consumed in India; however, typical preparation methods involve cooking, which kills the nematode larvae. In addition to studying rodent and mollusk hosts, monitor lizards can be used as effective sentinels. Sequence data are urgently needed to answer the question of the identity of *Angiostrongylus*-like metastrongylid nematodes isolated from all types of hosts. DNA-based diagnostic methods such as q-PCR and LAMP should be included in clinical diagnosis of suspected cases and in studies of genetic diversity and species identity of nematodes tentatively identified as *A. cantonensis*.

Keywords: *Angiostrongylus cantonensis; human angiostrongyliasis; Indian subcontinent;* eosinophilic meningitis

check for updates

Citation: Pandian, D.; Najer, T.; Modrý, D. An Overview of *Angiostrongylus cantonensis* (Nematoda: Angiostrongylidae), an Emerging Cause of Human Angiostrongylosis on the Indian Subcontinent. *Pathogens* **2023**, *12*, 851. https://doi.org/10.3390/pathogens12060851

Academic Editor: Anastasia Diakou

Received: 31 March 2023
Revised: 12 June 2023
Accepted: 13 June 2023
Published: 20 June 2023

1. Introduction

Angiostrongylus eosinophilic meningitis caused by *Angiostrongylus cantonensis* is a zoonotic disease that is currently spreading in the tropics and subtropics, with sporadic cases also occurring in temperate climatic zones [1,2]. The disease is caused by a metastrongyloid nematode, the rat lungworm, *Angiostrongylus cantonensis*. In humans, larval migration of this nematode typically results in eosinophilic meningitis and other central nervous system disorders. The increasing incidence and frequency of angiostrongylosis outbreaks has led to the disease being classified as an emerging infectious disease in Southeast Asian countries and in some invaded areas such as Australia and islands in the Pacific region.

Angiostrongylus cantonensis was first discovered in China (Chen, 1935), and it is primarily in Southeast Asia that is holding a tradition of its research in the Old World [2,3], reflecting the growing incidence of the disease [4]. Recently, infections of humans by the closely related *Angiostrongylus malaysiensis* (Bhaibulaya & Cross, 1971) have been reported in Southeast Asia [5], and the precise differentiation of the two nematode species deserves further attention, as does the possible neurotropism (the ability of larvae to specifically invade the brain and central nervous system) of other members of the genus.

The Indian subcontinent has never been systematically studied and reviewed for the presence of *A. cantonensis*. There, angiostrongylosis is known based on several human cases and limited field surveys [6]. Our aim of this review is to provide a thorough analysis of the available literature on human angiostrongylosis in the Indian subcontinent and contribute to the understanding of angiostrongylosis in the subcontinent. Data for this study were obtained by searching for the word combination "*Angiostrongylus cantonensis*" AND "*Angiostrongylus*" AND "*cantonensis*" AND "ocular angiostrongyliasis" AND ("India" OR ("Sri" AND "Lanka") OR "Pakistan" OR "Nepal" OR "Bhutan" OR " " Bangladesh" OR "Maldives") in the Web of Knowledge, Scopus, PubMed, and Google Scholar databases. This initial search revealed 23 relevant reports on angiostrongylosis in the selected geographic area. Then, references in all the studies obtained by the initial search were screened, and those with relevant information were added to the search results. Using this search method, 28 reports were included in our study, yielding 45 cases of angiostrongylosis in humans between 1966 and 2022. In the context of this study, we define the Indian subcontinent as Pakistan, India, Maldives, Nepal, Sri Lanka, Bangladesh, and Bhutan.

2. Biology and Life Cycle of *Angiostrongylus cantonensis*

A unique feature of the life cycle of *A. cantonensis* is its low host specificity at the intermediate host and paratenic host levels [1,7]. Natural definitive hosts include rats of the tribe Rattini (as defined by Lecompte et al.) [8]), with three species, *Rattus rattus* (Linnaeus, 1758), *Rattus norvegicus* (Berkenhout, 1769), and *Rattus exulans* (Peale, 1848), being the most reported definitive hosts [9,10]. Rats become infected by ingesting third-stage larvae (L3) in intermediate or paratenic hosts [2,11]. The ingested L3 penetrates the intestinal wall in the small intestine and enters the bloodstream. They then passively traverse the bloodstream and eventually reach the brain. Strong affinity toward the central nervous system (neurotropism) is the fundamental feature of the life cycle. In the central nervous system, L3 go through two molts and reaches the fifth-stage larvae L5 as immature adults in the subarachnoid space about two weeks after infection. Immature adults in typical definitive hosts continue to migrate to the terminal branches of the pulmonary arteries. There they become sexually mature, mate, and females lay eggs. The eggs hatch into first-stage larvae (L1) in the lung parenchyma, which ascend through the trachea and pharynx, then are swallowed into the digestive tract, and finally, the L1 leave the definitive hosts in feces. Snails or slugs many (= intermediate hosts) take up infected L1 from rat feces and these larvae develop into L3 [10,12,13]; typical developmental stages are depicted in Figure 1.

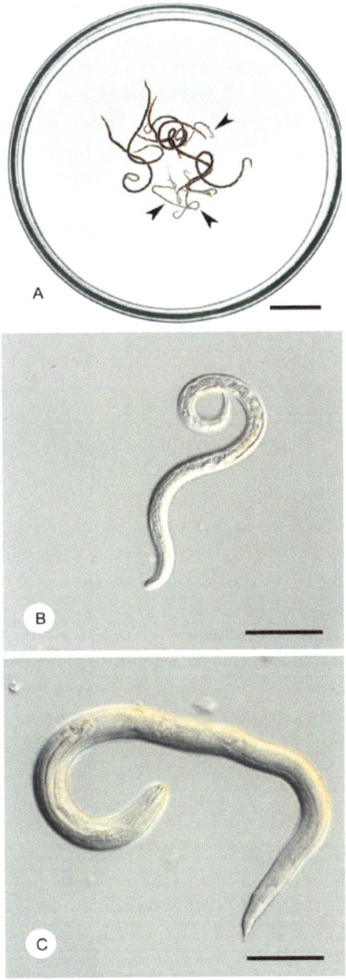

Figure 1. Developmental stages of *Angiostrongylus cantonensis* in intermediate and definitive hosts; Illustrative pictures depict the laboratory strain of *Angiostrongylus cantonensis* from Fatu Hiva, derived from snails from the Marquesas Islands, French Polynesia [14]. (**A**) Mixture of male and female adults removed from pulmonary arteries of a laboratory rat; females are larger (up to 30–35 mm), with typical barber-pole appearance, caused by the interweaving of the intestine and uterus; males (arrowheads) are smaller (max 15–25 mm), whitish, with well-developed bursa copulatrix; scale bar = 5 mm. (**B**) First-stage larva as shed in rat feces, scale bar = 50 μm. (**C**) Third-stage larva from a gastropod intermediate host, scale bar = 50 μm.

In addition to typical definitive hosts, gut-brain migration occurs in several aberrant hosts, including humans [2,15], domestic [16,17] and wild mammals [18], and birds [19]. However, L3 do not progress in these hosts and frequently cause eosinophilic meningitis accompanied by severe clinical symptoms, with dogs being the most common aberrant host other than humans. Multiple cases of canine angiostrongylosis in Australia show clinical manifestations of neurological deficits, cranial nerve dysfunction, fecal incontinence, hyperesthesia, seizures, ascending tail paresis, depression, diarrhea, and even death. There are also long-term neurological conditions, including tail paresis and hindlimb ataxia [20,21].

To date, the range of identified intermediate hosts of *A. cantonensis* is quite extensive and includes aquatic and terrestrial gastropods from at least 46 families [22]. Intermediate hosts become infected either by ingesting rat feces or by L1 actively penetrating their body wall or respiratory pores [23]. In intermediate hosts, L1 molt to L2 and then to L3, which persists in the tissues of infected mollusks. However, the L3 can leave the intermediate hosts spontaneously, either during its life or, usually, massively after its death. Outside of the host, L3 can survive in water for up to one month [24], with the ability to infect other intermediate hosts, further prolonging the survival of the larvae [25]. In addition to definitive hosts and intermediate hosts, L3 is also the stage that infects paratenic hosts and dead-end hosts and is the key stage of the cycle from an epidemiological perspective. The range of paratenic hosts includes both invertebrates (crustaceans, centipedes) and vertebrates (fish, amphibians, reptiles) [7], with many others likely still unknown.

3. Epidemiology and Clinical Manifestations of Human Angiostrongylosis in the Indian Subcontinent

Humans are a typical dead-end aberrant host. Angiostrongylosis caused by *A. cantonensis* was reported in the Indian subcontinent in both clinical forms as eosinophilic meningitis and/or ocular angiostrongylosis [26]. Generally, humans become infected by ingestion of L3 from tissues of intermediate or paratenic host, contaminated water, or vegetables [24]. The larvae penetrate the intestinal wall and, as they migrate, can cause inflammatory reactions in the organs they pass through [2]. *Angiostrongylus* eosinophilic meningitis (AEM) is the typical syndrome that occurs in humans and other aberrant vertebrate hosts [27]. When L3 reach the CNS, the larvae elicit clinical symptoms due to the direct destruction of nervous tissue, the consequent increase in intracranial pressure, and the host inflammatory response (which might get even higher in the event of larval death) [28,29]. The most common signs are severe headache, vomiting, fever, nausea, neck stiffness, paresthesia, hyperesthesia, and visual dystopia [30]. In children, fever, nausea, vomiting, somnolence, and constipation are more common than in adults [2]. AEM may eventually develop into encephalitis (especially in children), followed by coma and death [30].

Ocular angiostrongylosis appears to be more common in Asia than in the rest of *A. cantonensis* distributional range. A recent review [31] shows that most cases occur in a triangle with India to the west, Okinawa to the northeast, and New Guinea to the southeast. Outside this area, only three records are known, namely in South Africa [32], Jamaica [33], and Hawaii [34], all of them published after 2000. In most instances, ocular lesions are caused by a single nematode invading one eye along the optic nerve [26]. The worm is detected mainly in the anterior chamber or vitreous fluid without damaging the retina [31]; however, the subretinal space is also a common site of infection [35]. Patients often report blurred or floated vision [31]. Other series of ocular symptoms, including fundus changes, eye redness and pain, and progressive vision loss, eventually leading to blindness from patients, are also described by Dio et al. [26]. Optic neuritis may develop as a rare complication [31]. Treatment can be delivered in a variety of ways (e.g., surgery, laser, corticosteroids); it usually does not significantly improve visual outcome and focuses on preventing further damage caused by the parasite [35]. Co-occurrence of ocular form and AEM was recorded in 12 of 42 patients, summarized by Feng et al. [31]; only a single such case has been recorded in the Indian subcontinent [36]. In cases of ocular diseases, the nematodes recovered from human eyes were not molecularly characterized, leaving their identification uncertain.

Ocular angiostrongylosis represents an unusually high percentage of human *A. cantonensis* infections in the Indian subcontinent (Table 1), with 13 of 44 (29.5%) published cases from the subcontinent having ocular symptoms. This is even more evident in Sri Lanka, where 5 out of 7 (71.4%) of the ocular cases occur. In comparison, reports from Thailand and China estimate the prevalence of ocular cases to be around 1% [26,35,37]. The difference in the prevalence of ocular angiostrongylosis may suggest that an enormous number of AEM cases remain undiagnosed or unpublished or that *Angiostrongylus* migration within aberrant hosts in the Indian subcontinent differs from those in other parts of the world. The distribution of reports suggests that there may be many undiagnosed or unpublished cases (Figure 2). All published cases are from large metropolitan areas (e.g., Mumbai, Kottayam, Chennai, New Delhi, Colombo) with large hospitals that can diagnose and report the disease to a scientific community. In addition to the geographic pattern, records are also accumulating temporally (e.g., both reports from Madhya Pradesh are from 2019; [38,39], suggesting that the diagnosis of a case attracts attention and sparks eagerness to find more cases. In addition, rat and snail investigations are usually conducted after human cases have been discovered [40,41]. Regarding the age distribution of the patients, the ocular and AEM rate in children seems to be like that in adults [38,42–48], the small number of pediatric cases does not provide clear evidence. Notably, many case reports associate the infection with the consumption of monitor lizards; however, there are no data on the prevalence of *A. cantonensis* infection in saurian reptiles in the subcontinent (see below).

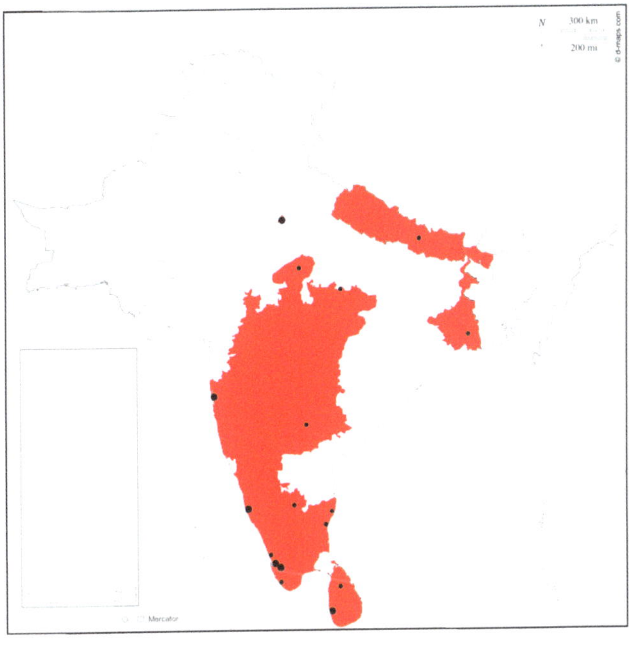

Figure 2. Schematic distribution of *Angiostrongylus cantonensis* reports on the Indian subcontinent. The red areas indicate countries or states with published records of angiostrongylosis, the black dots show cases—large dots multiple cases, and small dots single cases. Exact location of one Nepalese case is unknown, therefore, it is not indicated. The borders between Indian states are not shown, the map background was downloaded from https://d-maps.com, accessed on 8 November 2022.

Accurate and prompt diagnosis of eosinophilic meningitis caused by *Angiostrongylus cantonensis* is critical for appropriate treatment and prevention of sequelae. In general, systemic clinical examination, laboratory tests, and imaging studies can suggest meningitis caused by *A. cantonensis* in endemic areas, however, cannot provide final confirmation of an etiological agent. Elevated count of eosinophils in cerebrospinal fluid (CSF) is an important diagnostic feature and often the first step in the diagnostic process [2,28,49,50].

AEM can be diagnosed by immunodiagnostic methods using purified antigens or monoclonal antibodies in CSF or serum. Numerous techniques, including ELISA, immunoblot assays, gold immunochromatography, and rapid dot immunogold filtration assays, have been used for decades. However, these techniques may have limitations as the antibody may be undetectable in the early stages of infection [3,30,51,52]. Molecular diagnostic methods have been employed as a robust diagnostic tool based on PCR, real-time PCR, LAMP, and recombinase polymerase assay (RPA), which has the potential to improve AEM diagnosis by enabling highly sensitive detection of *A. cantonensis* DNA in a patient's CSF, serum, or other materials [51–61].

In the Indian subcontinent, 45 previously reported AEM cases were diagnosed based on factors such as a history of eating raw monitor lizard meat, clinical examination, laboratory findings (elevated eosinophils in blood or CSF), ophthalmologic examination and imaging techniques [48,62–67], supported by morphological identification of nematodes in cases when they were retrieved. Immunoblotting and qPCR were used to confirm AEM in two cases only [46,63], Table 1.

Table 1. Overview of reports of human angiostrongylosis on the Indian subcontinent, showing clinical form and diagnostic methods involved; definitive and intermediate hosts' data are included for those studies where they were associated with the investigation of human cases.

Country: State	EM Cases	OC Cases	Diagnostics Method	Monitor Lizard Consumption	References
India: Delhi		1	Slit lamp examination		[43]
India: Delhi		1	Clinical examination, ultrasound B scan, and fundus fluorescein angiography		[42]
India: Karnataka	1		History of raw monitor lizard meat consumption, clinical presentation, and EE in CSF	+	[64]
India: Karnataka	1		Serum and CSF antibodies to *A. cantonensis* 31-kDa antigen positive and magnetic resonance imaging (MRI)	+	[63]
India: Karnataka	1		MRI findings, CSF examination, larvae in CSF wet mount	+	[44]
India: Kerala	5		History of raw monitor lizard meat consumption, clinical presentation, and EE in CSF	+	[66]
India: Kerala	10		History of raw monitor lizard meat consumption, clinical presentation, EE in CSF., larvae in CSF wet mount, MRI	+	[48]
India: Kerala	3		EE in CSF and peripheral blood		[62]
India: Kerala	1 *		Ophthalmological examination, EE in CSF, and microscopic examination of the worm retrieved from eye		[36]
India: Kerala	1		EE in CSF, real-time PCR for *A. cantonensis*		[46]
India: Maharashtra	1		Not mentioned		[6]
India: Maharashtra	2		History of raw slug consumption and EE in CSF		[68]
India: Maharashtra	2		EE in CSF and, peripheral blood	**	[47]
India: Maharashtra	1		Histological examination of brain tissue after autopsy		[65]
India: Madhya Pradesh		1	Slit lamp examination and anterior segment optical coherence topography (AS-OCT)		[38]
India: Madhya Pradesh		1	Slit lamp examination and microscopic examination of the worm retrieved from eye		[39]
India: Puducherry	1		History of raw monitor lizard meat consumption, clinical presentation, EE in CSF analysis	+	[67]
India: Telangana	1		Computerized tomography (CT) scans, parietal craniotomy examination of the worm retrieved from cerebral abscess		[69]
India: West Bengal		1	Microscopic examination of the worm retrieved from eye		[6]

Table 1. *Cont.*

Country: State	EM Cases	OC Cases	Diagnostics Method	Monitor Lizard Consumption	References
Sri Lanka		1	Ophthalmological examination and microscopic examination of the worm retrieved from eye		[70,71]
Sri Lanka		1	Ophthalmological examination and microscopic examination of the worm retrieved from eye		[72]
Sri Lanka		1	Fundoscopic ophthalmological examination and microscopic examination of the worm retrieved from eye		[73]
Sri Lanka	1		EE in CSF and peripheral blood		[74]
Sri Lanka		1	Ophthalmological examination		[75]
Sri Lanka	1		History of raw monitor lizard meat consumption, clinical presentation, EE in CSF	+	[45]
Sri Lanka		1	Fundoscopic ophthalmological examination and microscopic examination of the worm retrieved from eye		[76]
Nepal: Kathmandu		1	Slit lamp examination, microscopic examination of the worm retrieved from eye		[77]
Nepal		1	Ophthalmological examination and microscopic examination of the worm retrieved from eye		[78]

Clinical form abbreviations and symbol definition: OC—ocular angiostrongylosis; EE—eosinophil examination; EM—eosinophilic meningitis; *—co-occurrence of OC and EM was reported by Baheti et al. [36]. +—Cases with preceding monitor lizard meat consumption. **—Reported contact with monitor lizard but not its consumption. Note: There is no single direct record of the presence of *Angiostrongylus* sp. in monitor lizard published from the subcontinent. Thus, all human cases associated with monitor meat consumption (Table 1) are only assumptions made from patients' anamneses.

4. Global Distribution and Prevalence of *Angiostrongylus cantonensis* in the Indian Subcontinent

A. cantonensis first became known in China in the 1930s and was observed there frequently on different hosts [10,12,79,80]. After the 1950s, numerous studies demonstrated the presence of *A. cantonensis* on various islands in the Pacific [10,12,81,82] and Oceanic regions [13], islands of the Indian Ocean [83]. In the late 1970s, the parasite was discovered in northeastern Africa [84]. In the New World, *A. cantonensis* has been found in some Caribbean islands, namely Cuba, the Dominican Republic, Jamaica, and Puerto Rico [85,86], and in continental Americas [87]. In recent years, it has become apparent that *A. cantonensis* is spreading at an alarming rate. Recent discoveries have been from South America and Brazil [88], the Canary Islands (Spain) [89], Mallorca (Balearic Islands, Spain) [90], Uganda [91], and North America [92,93]. The life cycle of *A. cantonensis* is typically associated with invasive definitive hosts and intermediate hosts, making it a textbook example of a multiple biological invasion. The African giant snail *Lissachatina fulica* (Bowdich, 1822) may have transmitted the parasite to the Pacific basin, where it spread rapidly within pre-existing rat populations [12]. It has been speculated that human activities such as global transport early in World War II may have contributed to the rapid spread of *A. cantonensis* [10,82].

In the Indian subcontinent, the presence of *A. cantonensis* was first detected in Sri Lanka during a survey by Alicata (1965). However, the first reported human case in Sri Lanka most likely dates from 1925 [70,71], i.e., before the parasite was formally described. If the determination is correct, the 1925 study is the first record of nematodes of the genus *Angiostrongylus* in a human host [94]. Later, *A. cantonensis* was also recorded in India [6]. In 1982, the first intermediate host survey took place in the Indian state of Maharashtra and confirmed the presence of *A. cantonensis* throughout India [40]. To date, the parasite has been detected in Sri Lanka, nine Indian states, and Nepal (see Table 1 and Figure 2). Most records (45 of 32 publications) are human clinical cases, supplemented by a few studies on definitive hosts [6,10,40,41,68,95] and intermediate hosts [40,68,96,97]. Paratenic hosts and incidental hosts other than humans have never been studied in the subcontinent.

5. *Angiostrongylus cantonensis* in the Definitive Rodent Host

Angiostrongylus cantonensis is thought to be largely associated with three invasive species of Rattini: *R. rattus*, *R. norvegicus*, and *R. exulans*, with local involvement of a few other rodent hosts [58]. The frequency with which infection spreads to other rodent species is largely unknown. An infection of *Sigmodon hispidus* (Say & Ord, 1825), a rodent host in the rather distant family Cricetidae, has been reported in North America [98].

Current knowledge about the distribution of species of the Rattini in the Indian subcontinent is very inconsistent. Most data relate to a few highly adaptable synanthropic species, while most taxa are endemic rodents with a virtually unknown natural history. According to the comprehensive concept of Rattini (as defined by Lecompte et al.) [8], 32 species of 10 genera (*Bandicota, Berylmys, Chiropodomys, Dacnomys, Leopoldamys, Micromys, Nesokia, Niviventer, Rattus, Vandeleuria*) of rats inhabit the Indian subcontinent [99–102]. The highest rat diversity occurs in the northeast of the subcontinent (e.g., 15 species in West Bengal, Figure 3), where areas overlap with several species from Southeast Asia (including the *A. cantonensis*). This is followed by Sri Lanka, the Western Ghats, and the Andaman and Nicobar Islands (8–9 species each); the diversity there is due to a high degree of local endemism [99]. Although many *A. cantonensis* records are known from the Western Ghats and Sri Lanka (Figure 2, Table 1), the involvement of endemic species in the life cycle of this parasite has never been studied.

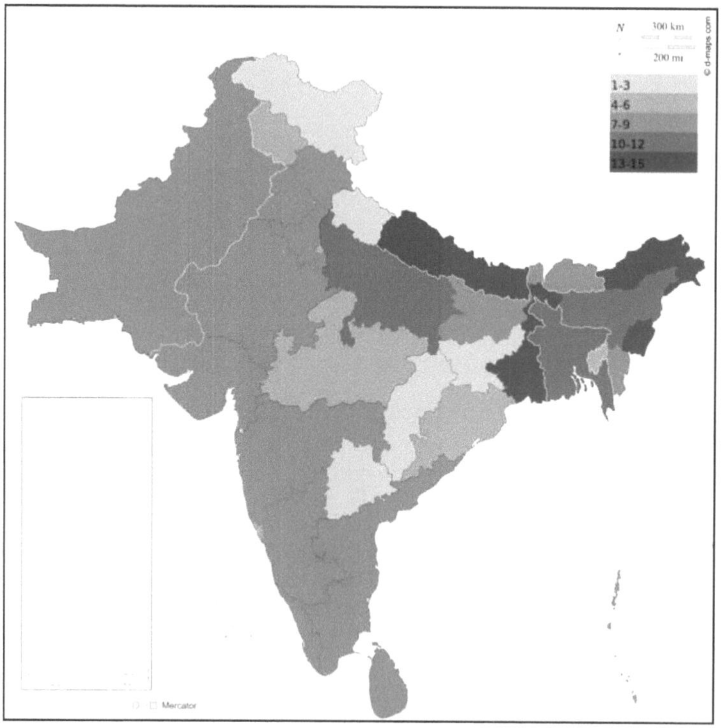

Figure 3. Diversity of rat species (Rattini) on the Indian subcontinent. The grayscale corresponds to the number of rat species described from each country or state, as indicated in the figure. Borders between countries white; coast, borders between Indian states, and outline of the subcontinent black. The map background was downloaded from https://d-maps.com, accessed on 8 November 2022.

Six species of rats inhabiting the Indian subcontinent were confirmed as *A. cantonensis* definitive hosts. Three species (*Bandicota indica, R. rattus, R. norvegicus*) are reported as hosts in studies directly from the subcontinent (Table 2) [6,10,40,41,83,95,96,103], while three others (*Berylmys bowersi, Niviventer fulvescens, Rattus exulans*) are known hosts in different parts of their distribution range [104]. From an ecological perspective, four of these species (*B. indica* (Bechstein, 1800), *R. rattus, R. norvegicus, R. exulans*) are synanthropic pests that frequently encounter humans [102]. The other species *B. bowersi* (Anderson, 1879), and *N. fulvescens* (Gray, 1847) [104], avoid human settlements. From the proven definitive hosts, *R. rattus* probably represents a major source of *A. cantonensis* infections and should be investigated; *R. exulans* is of minor importance due to its limited range in the subcontinent; *R. norvegicus* is typically found in large urban areas and seems unlikely to spread infection in rural areas [99]. On the other hand, data are lacking for several other synanthropic species. A total of 437 *Bandicota bengalensis* (Gray & Hardwicke, 1833) were examined by Alicata, Renapurkar et al. [40,83], and Limaye et al. [95], with no single *A. cantonensis* record. According to Agrawal (2000), this species displaces *R. norvegicus* in large urban areas, especially in Kolkata. If there is a difference in host competence between *B. bengalensis* and *R. norvegicus*, this could be the theoretical reason why only one human case is known from Kolkata, compared to Mumbai or Delhi (Table 1). *Rattus tanezumi* (Temminck, 1845) has only recently been separated from *R. rattus* [105,106], so in the case of *R. rattus* records, it cannot be clearly determined which species was examined. From a geographic perspective, *A. cantonensis* in rats was never surveyed in most of the subcontinent. The most conspicuous areas for further study are in the northeast of the subcontinent and associated islands (e.g., Andaman and Nicobar Islands, *R. rattus* was introduced in the Maldives [107]. (Figure 3). In general, the gaps in knowledge about the definitive hosts of *A. cantonensis* in the Indian subcontinent are compelling, considering that *A. cantonensis* is easily diagnosed and mainly associated with rats.

Table 2. List of records of *A. cantonensis* from hosts other than humans, as published from the Indian Subcontinent.

Country	State	*Bandicota indica* (Rodentia: Muridae)	*Rattus norvegicus* (Rodentia: Muridae)	*Laevicaulis alte* (Gastropoda: Veronicellidae)	*Macrochlamys indica* (Gastropoda: Ariophantidae)	Reference
India	Kerala	+				[41]
India	Maharashtra		+	+	+	[95,96]
India	Tamil nandu	+				[10,83]
Sri Lanka	Ceylon	+	+			[10,83]

"+"—Definitive and intermediate hosts were investigated in the Indian subcontinent by region.

5.1. *Angiostrongylus cantonensis in Intermediate Hosts*

Like most other metastrongylids, the life cycle of *A. cantonensis* invariably involves mollusks as obligate intermediate hosts. However, the nematode can develop in a wide range of gastropods, with extreme variation in prevalence among different populations [10,22,108]. Environmental factors, rat density, and the ecology of specific snail or slug species are likely responsible for the observed differences [22,58,82]. Importantly, *A. cantonensis* exploits both aquatic and terrestrial mollusks, which is one of the reasons for the differences in the local epidemiology of human infections [109,110]. As for gastropods in the Indian subcontinent, Tripathy and Mukhopadhyay (2015) provided a list of the freshwater mollusks of India [111], and Sen et al. summarized the diversity of terrestrial snails in India [112]. Their conclusion that there are 1129 species of terrestrial snails in India alone shows how difficult it is to grasp an enormous diversity of these invertebrates in the Indomalayan region. In addition, there are several smaller studies that list gastropods from different geographic or ecological parts of the subcontinent, such as mangrove mollusks from India [113] or terrestrial snails from Sri Lanka [114]. Many others also attempt to characterize diversity without providing indicative lists [115,116]. Given the low host specificity so far known in

A. cantonensis, it is easy to imagine that virtually any of these species could play the role of an *A. cantonensis* intermediate host.

In most studies, invasive snail species are considered more important than native fauna due to their ecology and high population density. The spread of *Lissachatina fulica*, one of the most detrimental invasive mollusk species, is commonly referred to as the gateway for the global spread of *A. cantonensis* [117,118]. *Bradybaena similaris* (A. Férussac, 1822), *Cornu aspersum* (O. F. Müller, 1774), *Parmarion martensi* (Simroth, 1893), *Pila* spp., *Pomacea canaliculata* (Lamarck, 1822), and *P. maculata* are associated with *A. cantonensis* in Southeast Asian countries, Australia, and the Caribbean islands [58,119–125]. Barrat et al. [58] provide a detailed overview of the prevalence and intensity of infection in species where they are known.

L. fulica and *P. canaliculata* are described as invasive in the Indian subcontinent [126–128], along with *Laevicaulis alte* (Férussac, 1822), *Physa acuta* (Draparnaud, 1805), and several other species [128–133]. *L. fulica* is common in almost all states, locally at densities, with negative impacts on agriculture [134]. *P. canaliculata* has invaded various water bodies in the Indian subcontinent [126], *L. alte* is widely reported in India and is known to have negative impacts on native snail species in the area [135]. Although there is no comprehensive study summarizing mollusk invasion across the subcontinent, the online data (www.iNaturalist.org, accessed on 8 November 2022) show a wide occurrence of the major invasive snails and slugs in India.

To date, few studies have addressed *A. cantonensis* in mollusks in the Indian subcontinent. Limaye et al. reported *A. cantonensis* infection in *Macrochlamys indica* (Godwin-Austen, 1883) [95]; the other few studies in the subcontinent [40,68,96,97] focused on a single species, the invasive slug *L. alte* [68].

5.2. Snail Consumption

Limited information is available on the scale of edible snail consumption in the Indian subcontinent. Snail consumption is well-known in some parts of India, such as the northeastern region, West Bengal, and other places such as Bihar, Karnataka, Kerala, Madhya Pradesh, and Tamil Nadu [136–139]. In these regions, snail meat is well known among urbanites and rural tribal communities for its therapeutic and culinary uses [140]. Although the Indian Council of Agricultural Research (ICAR) has supported the introduction of snail farming, there are few snail farms in the country. Instead, snails are collected from the wild rather than being cultivated [141]. Supplementary Table S1 provides an overview of mollusk consumption. Freshwater snails, *Pila globosa* (Swainson, 1822), *Bellamya bengalensis* (Lamarck, 1822), *Viviparus viviparus* (Linnaeus, 1758), and several species of terrestrial snails are among the snails reported to be most consumed in many parts of India [140,142–144]. Sharma et al. reported a case of angiostrongylosis in humans after consumption of raw slugs *L. alte* [68] but this species has not been mentioned in studies on the consumption of edible mollusks.

Consumption of raw or insufficiently cooked snails is a common source of human infection in Southeast Asian countries such as China, Taiwan, Thailand, and Hong Kong, including reports of associated clusters of infection [109,145]. However, nematode larvae are sensitive to high temperatures, and even short boiling kills L3 of *A. cantonensis* in infected mollusks [146]. Snails used in reviewed traditional Indian dishes are always prepared by boiling or frying for 5–10 min with various flavors and spices. Technically, following these procedures prevents the presence of live infectious larvae in cooked dishes. Importantly, many recipes recommend soaking the snails in water for 24 h before use. Together with the initial cleaning, this is a critical moment that deserves attention from an epidemiological point of view. The L3 actively escape from snails [147] and can contaminate cooking surfaces and utensils in high numbers [24]. Reportedly, the water from the soaked snails is used as eye drops to treat conjunctivitis as a traditional remedy [148], which may pose an additional risk of angiostrongylosis since larvae may enter the digestive system through the nasolacrimal duct.

6. *Angiostrongylus cantonensis* in Paratenic Reptilian and Amphibian Hosts in the Context of Local Consumption in India

6.1. Monitor Lizards (Varanidae)

Both amphibians and reptiles have been identified as natural paratenic hosts of *A. cantonensis* [7,149]. In this context, monitor lizards (Varanidae) are most mentioned; several case reports from the Indian subcontinent include a history of monitor lizard consumption preceding AEM symptoms [44,45,48,63,64,66,67]. Although the taxonomy of monitor lizards has not been clearly established [150], four ecologically distinct species inhabit the subcontinent: *Varanus bengalensis* (Daudin, 1802), *V. flavescens* (Hardwicke & Gray, 1827), *V. griseus* (Daudin, 1803), and *V. salvator* (Laurenti, 1768). As potential *A. cantonensis* paratenic hosts, *V. griseus* and *V. flavescens* can be excluded based on their ecology or rarity [151]. In the reports associating *A. cantonensis* infection with monitor lizard meat consumption, species identification was not discussed, and all cases were automatically assigned to the Bengal monitor lizard—*V. bengalensis*. It is the most abundant species throughout the subcontinent [152], a terrestrial animal inhabiting a wide range of habitats from tropical to temperate [153], and a documented paratenic host of *A. cantonensis* in Thailand [149]; the species has been shown to feed on mollusks. The water monitor, *V. salvator*, could be a second potentially important paratenic host. It is a semiaquatic or amphibious species that lives near bodies of water [154] and has been shown to cross the sea between islands. However, it can also live on land and adapt to a variety of food sources, such as human waste when it seasonally visits tourist sites [151]. Because of its ability to disperse to islands, it is also the only monitor species in the Andaman and Nicobar Islands [152]. Cannibalism is known in both monitor species [151], but its role in the life cycle of *A. cantonensis* remains to be investigated. Despite the above, the assumption that consumption of raw monitor meat is a major cause of angiostrongylosis in humans does not seem to be based so much on facts. *V. bengalensis* was found to be a paratenic host harboring L3 of *A. cantonensis* in Thailand [149], and a subsequent study found 100% prevalence at five of four Thai sites [155]. However, these are the only two studies that directly detected *A. cantonensis* in monitor lizards. All subsequent records associate the cases only with the consumption of monitor lizards mentioned by the patients. In the Indian subcontinent, the presence of *A. cantonensis* in amphibians or reptiles has never been directly demonstrated in any study, but it is likely (see, e.g., Anettová et al.) [156]. Of six records in which patients admit to previous consumption of mnitor lizards [44,45,48,63,64,66,67], only one is described in detail, likely ruling out other possible sources of infection [45]. If all these associations were true, it would mean that 47.6% of human cases were caused by the consumption of reptiles—an unusually high number compared to more sporadic cases from SE Asia, e.g., Yang et al. [157]

6.2. Snakes

Little attention was paid to the role of snakes in the life cycle of *A. cantonensis* [158]. In India, the consumption of snake meat is not considered a delicacy. However, numerous articles describe snake meat, gallbladder, and skin used in traditional therapies by tribal communities [159–166]. There have been no published cases of neurological or ocular disease caused by *A. cantonensis* linked with the consumption of snake products. Depending on the region and tribe, their products are used in different ways; the most used snake species are listed in Supplementary Table S2.

6.3. Amphibians

The role of amphibians in the life cycle of *Angiostrongylus cantonensis* deserves more attention. Limited studies have experimentally demonstrated frogs as paratenic hosts [167], and *A. cantonensis* infection has also been detected in free-living frogs [158,168,169]. Reports of angiostrongylosis in humans due to the consumption of frog meat have come from China, Japan, Taiwan, and the United States [7,158,168]. Unlike monitor lizards, frogs are ignored as a source of *A. cantonensis* infection in the Indian subcontinent. In India, amphibians play an important role in cultural traditions, and tribal communities throughout the country rely on amphibians for a variety of uses, including food and traditional medicine [170,171]. However, frog meat is not commonly consumed in India [172]. Due to the trade ban on the export of frog meat and the restriction on the collection of frogs in India [173], they are not widely consumed in the market, yet frog meat continues to be consumed in tribal communities across the country [161,174–176]. The most popular way to consume frogs for medicinal purposes is to boil the meat and eat it dry fried or in soup [176]. The *Hoplobatrachus tigerinus* (Daudin, 1802), *Nasikabatrachus sahyadrensis* (Biju & Bossuyt, 2003), and *Euphlyctis cyanophlyctis* (Schneider, 1799) frogs are the most consumed frog species in India [175–177]. Supplementary Table S3 provides examples of frog meat consumption in India.

To our knowledge, frogs have not been investigated as potential hosts for *A. cantonensis* in India, and no cases have been reported associated with the consumption of frog meat, which contrasts with numerous human cases of angiostrongylosis associated with the consumption of raw monitor lizard meat. Unlike frogs, monitor lizards are large animals that are not consumed every day. It is possible that patients affected by angiostrongylosis are likely to remember eating them, associate these two events, and report this to their physician.

Importantly, most studies on the consumption of amphibians and reptiles in the Indian subcontinent report situations in which the meat of these animals is prepared in various ways, including frying, roasting, and boiling. This should be given more attention in the future because nematode larvae cannot survive in cooked animal meat; rather, the infectious L3 from *A. cantonensis* can contaminate kitchenware and cooking utensils and cause infections in humans.

7. Conclusions

Despite numerous cases of angiostrongylosis in humans, *A. cantonensis* research in the Indian subcontinent has lagged behind neighboring Southeast Asia and China. Published studies on the possible life cycle invariably refer to infections in humans. In terms of definitive hosts, *A. cantonensis* has been studied locally only in synanthropic rodents, making the potential sylvatic cycle an unexplored area for future research.

Geographically, *A. cantonensis* is well known in southern India and Sri Lanka, with most records coming from large metropolitan areas with large hospitals and dense human populations. Rat species richness appears to play a minor role, as hotspots of diversity do not overlap with the frequency of human cases. The very narrow focus of the published literature and the uneven geographic distribution suggests that *A. cantonensis* infections may be significantly underdiagnosed. The clinical symptoms divide human angiostrongylosis into two syndromes—eosinophilic meningitis and ocular angiostrongylosis. The separation of the two syndromes raises the question of whether they are caused by the same nematode species. Nematode identification was invariably based on nematode morphology, with no DNA sequences available from *A. cantonensis* throughout the subcontinent. Available molecular data [178–186] distinguish three related species, namely *A. cantonensis*, *A. malaysiensis*, and *A. mackerrasae* [183]. The possible presence of *A. malaysiensis* or some yet undescribed *Angiostrongylus* species and the hypothetical infection of humans by other metastrongyloid nematodes further complicates the question of the identity of *A. cantonensis* in the Indian subcontinent. Molecular studies that enable proper identification and description of genetic variability of *Angiostrongylus* in the Indian subcontinent are urgently needed, as is awareness and capacity building in the field of *A. cantonensis* diagnostics among medical professionals.

Supplementary Materials: The following supporting information can be downloaded at: https://www.mdpi.com/article/10.3390/pathogens12060851/s1, Table S1: List of Indian states where eating snail meat is common; Table S2: Overview of snakes are used for zootherapy in tribal communities; Table S3: Overview of where frog meat is common. References [174,187–211] are cited in the Supplementary Materials.

Author Contributions: Conceptualization by D.M. and D.P.; Writing and original draft preparation: D.M., D.P. and T.N. All authors have read and agreed to the published version of the manuscript.

Funding: The study was supported by Czech Science Foundation grant no. 22-26136S.

Data Availability Statement: Not applicable.

Acknowledgments: I would like to express my sincere gratitude to David Modrý and Tomáš Najer for their excellent assistance in revising this paper. Their insightful comments and suggestions greatly improved the quality of this article. I would also like to thank the Department of Veterinary Sciences at CZU for their valuable resources and supportive research environment. Also, I would like to thank Jignesh Italia from the Faculty of Tropical Science at CZU for his technical assistance.

Conflicts of Interest: The authors declare no conflict of interest.

References

1. Morgan, E.R.; Modry, D.; Paredes-Esquivel, C.; Foronda, P.; Traversa, D. Angiostrongylosis in animals and humans in Europe. *Pathogens* **2021**, *10*, 1236. [CrossRef] [PubMed]
2. Wang, Q.-P.; Lai, D.-H.; Zhu, X.-Q.; Chen, X.-G.; Lun, Z.-R. Human angiostrongyliasis. *Lancet Infect. Dis.* **2008**, *8*, 621–630. [CrossRef] [PubMed]
3. Eamsobhana, P. Angiostrongyliasis in Thailand: Epidemiology and laboratory investigations. *Hawaii J. Med. Public Health* **2013**, *72*, 28–32. [PubMed]
4. Punyagupta, S.; Bunnag, T.; Juttijudata, P.; Rosen, L. Eosinophilic meningitis in Thailand. Epidemiologic studies of 484 typical cases and the etiologic role of *Angiostrongylus cantonensis*. *Am. J. Trop. Med. Hyg.* **1970**, *19*, 950–958. [CrossRef] [PubMed]
5. Watthanakulpanich, D.; Jakkul, W.; Chanapromma, C.; Ketboonlue, T.; Dekumyoy, P.; Lv, Z.; Chan, A.H.E.; Thaenkham, U.; Chaisiri, K. Co-occurrence of *Angiostrongylus malaysiensis* and *Angiostrongylus cantonensis* DNA in cerebrospinal fluid: Evidence from human eosinophilic meningitis after ingestion of raw snail dish in Thailand. *Food Waterborne Parasitol.* **2021**, *24*, e00128. [CrossRef]
6. Parmeter, S.N.; Chowdhury, A.B. *Angiostrongylus cantonensis* in India. Bull. Calcutta Sch. *Trop. Med.* **1966**, *14*, 38.
7. Turck, H.C.; Fox, M.T.; Cowie, R.H. Paratenic hosts of *Angiostrongylus cantonensis* and their relation to human neuroangiostrongyliasis globally. *One Health* **2022**, *15*, 100426. [CrossRef]
8. Lecompte, E.; Aplin, K.; Denys, C.; Catzeflis, F.; Chades, M.; Chevret, P. Phylogeny and biogeography of African Murinae based on mitochondrial and nuclear gene sequences, with a new tribal classification of the subfamily. *BMC Evol. Biol.* **2008**, *8*, 199. [CrossRef]
9. Chen, H.-T. Un Nouveau nématode pulmonaire, *Pulmonema cantonensis*, n.g., n. sp. *Ann. Parasitol. Hum. Comp.* **1935**, *13*, 312–317. [CrossRef]
10. Alicata, J.E. Biology and distribution of the Rat lungworm, *Angiostrongylus cantonensis*, and its relationship to eosinophilic meningoencephalitis and other neurological disorders of man and animals. *Adv. Parasitol.* **1965**, *3*, 223–248. [CrossRef]
11. Walden, H.D.S.; Slapcinsky, J.D.; Roff, S.; Calle, J.M.; Goodwin, Z.D.; Stern, J.; Corlett, R.; Conway, J.; McIntosh, A. Geographic distribution of *Angiostrongylus cantonensis* in wild rats (*Rattus rattus*) and terrestrial snails in Florida, USA. *PLoS ONE* **2017**, *12*, e0177910. [CrossRef]
12. Wallace, G.D.; Rosen, L. Studies on eosinophilic meningitis: I. Observations on the geographic distribution of *Angiostrongylus cantonensis* in the Pacific area and its prevalence in wild rats. *Am. J. Epidemiol.* **1965**, *81*, 52–62. [CrossRef]
13. Mackerras, M.; Sandars, D. The life history of the rat lung-worm, *Angiostrongylus cantonensis* (Chen) (Nematoda: Metastrongylidae). *Aust. J. Zool.* **1955**, *3*, 1–21. [CrossRef]
14. Červená, B.; Modrý, D.; Fecková, B.; Hrazdilová, K.; Foronda, P.; Alonso, A.M.; Lee, R.; Walker, J.; Niebuhr, C.N.; Malik, R.; et al. Low diversity of *Angiostrongylus cantonensis* complete mitochondrial DNA sequences from Australia, Hawaii, French Polynesia and the Canary Islands revealed using whole genome next-generation sequencing. *Parasites Vectors* **2019**, *12*, 241. [CrossRef] [PubMed]
15. Alicata, J. The discovery of *Angiostrongylus cantonensis* as a cause of human eosinophilic meningitis. *Parasitol. Today* **1991**, *7*, 151–153. [CrossRef]
16. Mason, K.V.; Prescott, C.W.; Kelly, W.R.; Waddell, A.H. Granulomatous encephalomyelitis of puppies due to *Angiostrongylus cantonensis*. *Aust. Vet. J.* **1976**, *52*, 295. [CrossRef]
17. Wright, J.; Kelly, W.; Hamilton, J.; Wadell, A. Equine neural angiostrongylosis. *Aust. Veter J.* **1991**, *68*, 58–60. [CrossRef]
18. Carlisle, M.S.; Prociv, P.; Grennan, J.; Pass, M.A.; Campbell, G.L.; Mudie, A. Cerebrospinal angiostrongyliasis in five captive tamarins (*Sanguinus* spp.). *Aust. Veter J.* **1998**, *76*, 167–170. [CrossRef] [PubMed]
19. Barrett, J.; Carlisle, M.; Prociv, P. Neuro-angiostrongylosis in wild Black and Grey-headed flying foxes (*Pteropus* spp.). *Aust. Vet. J.* **2002**, *80*, 554–558. [CrossRef] [PubMed]
20. Mason, K.V. Canine neural angiostrongylosis: The clinical and therapeutic features of 55 natural cases. *Aust. Vet. J.* **1987**, *64*, 201–203. [CrossRef]

21. Lunn, J.A.; Lee, R.; Smaller, J.; MacKay, B.M.; King, T.; Hunt, G.B.; Martin, P.; Krockenberger, M.B.; Spielman, D.; Malik, R. Twenty two cases of canine neural angiostrongylosis in eastern Australia (2002–2005) and a review of the literature. *Parasites Vectors* **2012**, *5*, 70. [CrossRef]
22. Kim, J.R.; Hayes, K.A.; Yeung, N.W.; Cowie, R.H. Diverse gastropod hosts of *Angiostrongylus cantonensis*, the rat lungworm, globally and with a focus on the Hawaiian Islands. *PLoS ONE* **2014**, *9*, e94969. [CrossRef] [PubMed]
23. Thiengo, S.C.; Simões, R.D.O.; Fernandez, M.A.; Maldonado, A. *Angiostrongylus cantonensis* and rat lungworm disease in Brazil. *Hawaii J. Med. Public Health* **2013**, *72*, 18–22.
24. Howe, K.; Kaluna, L.; Lozano, A.; Fischer, B.T.; Tagami, Y.; McHugh, R.; Jarvi, S. Water transmission potential of *Angiostrongylus cantonensis*: Larval viability and effectiveness of rainwater catchment sediment filters. *PLoS ONE* **2019**, *14*, e0209813. [CrossRef]
25. Modrý, D.; Fecková, B.; Putnová, B.; Manalo, S.M.; Otranto, D. Alternative pathways in *Angiostrongylus cantonensis* (Metastrongyloidea: Angiostrongylidae) transmission. *Parasitology* **2021**, *148*, 167–173. [CrossRef] [PubMed]
26. Diao, Z.; Wang, J.; Qi, H.; Li, X.; Zheng, X.; Yin, C. Human ocular angiostrongyliasis: A literature review. *Trop. Doct.* **2011**, *41*, 76–78. [CrossRef] [PubMed]
27. Beaver, P.C.; Rosen, L. Memorandum on the first report of *Angiostrongylus* in man, by Nomura and Lin, 1945. *Am. J. Trop. Med. Hyg.* **1964**, *13*, 589–590. [CrossRef]
28. Ansdell, V.; Wattanagoon, Y. *Angiostrongylus cantonensis* in travelers: Clinical manifestations, diagnosis, and treatment. *Curr. Opin. Infect. Dis.* **2018**, *31*, 399–408. [CrossRef]
29. Kwon, E.; Ferguson, T.M.; Park, S.Y.; Manuzak, A.; Qvarnstrom, Y.; Morgan, S.; Ciminera, P.; Murphy, G.S. A severe case of *Angiostrongylus* eosinophilic meningitis with encephalitis and neurologic sequelae in Hawa'i. *Hawaii J. Med. Public Health* **2013**, *72*, 41–45.
30. Murphy, G.S.; Johnson, S. Clinical aspects of eosinophilic meningitis and meningoencephalitis caused by *Angiostrongylus cantonensis*, the rat lungworm. *Hawaii J. Med. Public Health* **2013**, *72*, 35.
31. Feng, Y.; Nawa, Y.; Sawanyavisuth, K.; Lv, Z.; Wu, Z.-D. Comprehensive review of ocular angiostrongyliasis with special reference to optic neuritis. *Korean J. Parasitol.* **2013**, *51*, 613–619. [CrossRef]
32. Kumar, V.; Kyprianou, I.; Keenan, J.M. Ocular angiostrongyliasis: Removal of a live nematode from the anterior chamber. *Eye* **2004**, *19*, 229–230. [CrossRef]
33. Mattis, A.; Mowatt, L.; Lue, A.; Lindo, J.; Vaughan, H. Ocular angiostrongyliasis—First case report from Jamaica. *West Indian Med. J.* **2009**, *58*, 383–385. [PubMed]
34. McCarthy, K.; Liu, K.; Kokame, G.T.; Merrill, P.T.; Gilca, M.; Cohen, J. First case of subretinal ocular angiostrongyliasis associated with retinal detachment in the United States. *Hawaii J. Health Soc. Welf.* **2021**, *80*, 40. [PubMed]
35. Sinawat, S.; Trisakul, T.; Choi, S.; Morley, M.; Sinawat, S.; Yospaiboon, Y. Ocular angiostrongyliasis in Thailand: A retrospective analysis over two decades. *Clin. Ophthalmol.* **2019**, *13*, 1027–1031. [CrossRef] [PubMed]
36. Baheti, N.N.; Sreedharan, M.; Krishnamoorthy, T.; Nair, M.D.; Radhakrishnan, K. Eosinophilic meningitis and an ocular worm in a patient from Kerala, South India. *Case Rep.* **2009**, *2009*, bcr2007122093. [CrossRef] [PubMed]
37. Sawanyawisuth, K.; Kitthaweesin, K.; Limpawattana, P.; Intapan, P.; Tiamkao, S.; Jitpimolmard, S.; Chotmongkol, V. Intraocular angiostrongyliasis: Clinical findings, treatments and outcomes. *Trans. R. Soc. Trop. Med. Hyg.* **2007**, *101*, 497–501. [CrossRef]
38. Sen, P.; Date, A.; Jain, E.; Mohan, A. *Angiostrongylus*—A technique for removing a rare parasite from the cornea of a child. *J. Am. Assoc. Pediatr. Ophthalmol. Strabismus* **2019**, *23*, 119–120. [CrossRef]
39. Aishwarya, A.; Tiwari, U.S.; Gandhi, S.; Sisodia, P. *Angiostrongylus cantonensis* in anterior chamber. *Indian J. Ophthalmol.* **2019**, *67*, 158–160. [CrossRef]
40. Renapurkar, D.M.; Bhopale, M.K.; Limaye, L.S.; Sharma, K.D. Prevalence of *Angiostrongylus cantonensis* infection in commensal rats in Bombay. *J. Helminthol.* **1982**, *56*, 345–349. [CrossRef]
41. Thomas, M.; Thangavel, M.; Thomas, R.P. *Angiostrongylus cantonensis* (Nematoda, Metastrongylidae) in bandicoot rats in Kerala, South India. *Infect. Dis.* **2015**, *8*, 324–326.
42. Mehta, D.K.; Arora, R.; Chauhan, D.; Shroff, D.; Narula, R. Chemo-paralysis for the removal of a live intraocular worm in ocular angiostrongyliasis. *Clin. Exp. Ophthalmol.* **2006**, *34*, 493–495. [CrossRef]
43. Malhotra, S.; Mehta, D.K.; Arora, R.; Chauhan, D.; Ray, S.; Jain, M. Ocular angiostrongyliasis in a child—First case report from India. *J. Trop. Pediatr.* **2006**, *52*, 223–225. [CrossRef]
44. Rai, S.; Madi, D.; Pai, S.; Baliga, S. Unusual larva in the CSF and unique MRI findings in a case of eosinophilic meningitis. *J. Clin. Imaging Sci.* **2014**, *4*, 76. [CrossRef]
45. Hidelaratchi, M.D.P.; Riffsy, M.T.M.; Wijesekera, J.C. A case of eosinophilic meningitis following monitor lizard meat consumption, exacerbated by anthelminthics. *Ceylon Med. J.* **2010**, *50*, 84–86. [CrossRef]
46. Raveendrakumar, A.G.; Remadevi, G.S.; Eapen, E.K.; Vijayamma, A.K.T. Eosinophilic meningitis in a toddler. *Indian Pediatr.* **2021**, *58*, 187–188. [CrossRef]
47. Shah, I.; Barot, S.; Madvariya, M. Eosinophilic meningitis: A case series and review of literature of *Angiostrongylus cantonensis* and *Gnathostoma spinigerum*. *Indian J. Med. Microbiol.* **2015**, *33*, 154–158. [CrossRef] [PubMed]
48. Parameswaran, K. Case series of eosinophilic meningoencephalitis from South India. *Ann. Indian Acad. Neurol.* **2006**, *9*, 217. [CrossRef]
49. Ianas, V.; Kuberski, T. Eosinophilic meningitis due to *Angiostrongylus cantonensis*. *Infect. Med.* **2008**, *25*, 366–368.

50. Wang, Q.-P.; Wu, Z.-D.; Wei, J.; Owen, R.L.; Lun, Z.-R. Human *Angiostrongylus cantonensis*: An update. *Eur. J. Clin. Microbiol. Infect. Dis.* **2012**, *31*, 389–395. [CrossRef]
51. Wilkins, P.P.; Qvarnstrom, Y.; Whelen, A.C.; Saucier, C.; Da Silva, A.J.; Eamsobhana, P. The current status of laboratory diagnosis of *Angiostrongylus cantonensis* infections in humans using serologic and molecular methods. *Hawaii J. Med. Public Health* **2013**, *72*, 55–57.
52. de Melo, L.C.V.; de Souza, F.C.R.; Baccin, A.D.O.; Mota, D.J.G.D.; Pereira-Chioccola, V.L.; Pinto, P.L.S. Immunoanalysis of different antigenic preparations of *Angiostrongylus cantonensis* for neuroangiostrongyliasis diagnosis improvement. *Mem. Inst. Oswaldo Cruz* **2022**, *117*, e220086. [CrossRef]
53. Qvarnstrom, Y.; Sullivan, J.J.; Bishop, H.S.; Hollingsworth, R.; Da Silva, A.J. PCR-based detection of *Angiostrongylus cantonensis* in tissue and mucus secretions from molluscan hosts. *Appl. Environ. Microbiol.* **2007**, *73*, 1415–1419. [CrossRef] [PubMed]
54. Qvarnstrom, Y.; Xayavong, M.; Chea, N.; da Silva, A.C.A.; Heng, S.; Park, S.Y.; Calimlim, P.S.; Johnson, S.; Whelen, A.C.; Fox, L.M.; et al. Real-time polymerase chain reaction detection of *Angiostrongylus cantonensis* DNA in cerebrospinal fluid from patients with eosinophilic meningitis. *Am. J. Trop. Med. Hyg.* **2016**, *94*, 176–181. [CrossRef]
55. Maleewong, W.; Sombatsawat, P.; Intapan, P.M.; Wongkham, C.; Chotmongkol, V. Immunoblot evaluation of the specificity of the 29-kDa antigen from young adult female worms *Angiostrongylus cantonensis* for immunodiagnosis of human angiostrongyliasis. *Asian Pac. J. Allergy Immunol.* **2001**, *19*, 267.
56. Eamsobhana, P.; Gan, X.; Ma, A.; Wang, Y.; Wanachiwanawin, D.; Yong, H. Dot immunogold filtration assay (DIGFA) for the rapid detection of specific antibodies against the rat lungworm *Angiostrongylus cantonensis* (Nematoda: Metastrongyloidea) using purified 31-kDa antigen. *J. Helminthol.* **2014**, *88*, 396–401. [CrossRef]
57. Chen, J.-X.; Chen, M.-X.; Ai, L.; Chen, J.-H.; Chen, S.-H.; Zhang, Y.-N.; Cai, Y.-C.; Zhu, X.-Q.; Zhou, X.-N. A protein microarray for the rapid screening of patients suspected of infection with various food-borne helminthiases. *PLoS Negl. Trop. Dis.* **2012**, *6*, e1899. [CrossRef] [PubMed]
58. Barratt, J.; Chan, D.; Sandaradura, I.; Malik, R.; Spielman, D.; Lee, R.; Marriott, D.; Harkness, J.; Ellis, J.; Stark, D. *Angiostrongylus cantonensis*: A review of its distribution, molecular biology and clinical significance as a human pathogen. *Parasitology* **2016**, *143*, 1087–1118. [CrossRef] [PubMed]
59. Liu, C.-Y.; Song, H.-Q.; Zhang, R.-L.; Chen, M.-X.; Xu, M.-J.; Ai, L.; Chen, X.-G.; Zhan, X.-M.; Liang, S.-H.; Yuan, Z.-G.; et al. Specific detection of *Angiostrongylus cantonensis* in the snail *Achatina fulica* using a loop-mediated isothermal amplification (LAMP) assay. *Mol. Cell. Probes* **2011**, *25*, 164–167. [CrossRef]
60. Sears, W.J.; Qvarnstrom, Y.; Nutman, T.B. RPAcan3990: An ultrasensitive recombinase polymerase assay to detect *Angiostrongylus cantonensis* DNA. *J. Clin. Microbiol.* **2021**, *59*, e01185-21. [CrossRef]
61. Sears, W.J.; Qvarnstrom, Y.; Dahlstrom, E.; Snook, K.; Kaluna, L.; Baláž, V.; Feckova, B.; Šlapeta, J.; Modry, D.; Jarvi, S. AcanR3990 qPCR: A novel, highly sensitive, bioinformatically-informed assay to detect *Angiostrongylus cantonensis* Infections. *Clin. Infect. Dis.* **2021**, *73*, e1594–e1600. [CrossRef]
62. Jalal, M.J.A.; Varghese, B.; Menon, M.K.; Pratap, T.; Moheen, A.; Mathew, J.; Vijayan, T.P.; Tomy, M.P. Eosinophilic meningitis: Report of three cases from a hilly area in Kerala. *Ann. Indian Acad. Neurol.* **2019**, *22*, 108–111. [CrossRef]
63. Nalini, A.; Ramakrishna, A.; Dekumoy, P.; Kumar, R.; Pakdee, W.; Saini, J.; Hegde, V. Severe form of radiculo-myelo-neuropathy with meningo-encephalitis secondary to *Angiostrongylus cantonensis* infection: Unusual corpus callosal lesions and serial magnetic resonance imaging findings. *Neurol. India* **2013**, *61*, 414–418. [CrossRef]
64. Pai, S.; Madi, D.; Achappa, B.; Mahalingam, S.; Kendambadi, R. An interesting case of eosinophilic meningitis. *J. Clin. Diagn. Res.* **2013**, *7*, 734. [CrossRef] [PubMed]
65. Chaudhari, J.P.; Shenoy, A.S.; Goel, N.A. Eosinophilic meningitis due to *A. Cantonensis* revealed at autopsy. *Indian J. Pathol. Microbiol.* **2022**, *65*, 420. [PubMed]
66. Panackel, C.; Vishad; Cherian, G.; Vijayakumar, K.; Sharma, R. Eosinophilic meningitis due to *Angiostrongylus cantonensis*. *Indian J. Med. Microbiol.* **2006**, *24*, 220–221. [CrossRef] [PubMed]
67. Hamide, A.; Johny, J.; Kumar, A.; Kolar, V.V. Eosinophilic meningitis caused by consumption of meat of monitor lizard (*Varanus bengalensis*). *Neurol. India* **2018**, *66*, 1166. [CrossRef] [PubMed]
68. Sharma, K.D.; Renapurkar, D.M.; Bhopale, M.K.; Nathan, J.; Boraskar, A.; Chotani, S. Study of a focus of *Angiostrongylus cantonensis* in Greater Bombay. *Bull Haffkine* **1981**, *9*, 38–46.
69. Purohit, A.K.; Dinakar, I.; Sundaram, C.; Ratnakar, K.S. *Angiostrongylus cantonensis* abscess in the brain. *J. Neurol. Neurosurg. Psychiatry* **1991**, *54*, 1015–1016. [CrossRef] [PubMed]
70. Joseph, H.P. Clinical notes on the effects of a nematode larva in the eye. *Ceylon J. Sci. Med. Sci.* **1927**, *1*, 141.
71. Nicholls, L. Note on a nematode larva removed from the eye of a man. *Ceylon J. Sci.* **1925**, *1*, 145.
72. Durette-Desset, M.-C.; Chabaud, A.G.; Cassim, M.H.S.; Ismail, M.M.; Premaratne, U.N.; Abeyewickreme, W.; Dissanaike, A.S. On an infection of a human eye with Parastrongylus (=Angiostrongylus) sp. in Sri Lanka. *J. Helminthol.* **1993**, *67*, 69–72. [CrossRef] [PubMed]
73. Wariyapola, D.; Goonesinghe, N.; Priyamanna, T.H.; Fonseka, C.; Ismail, M.; Abeyewickreme, W.; Dissanaike, A. Second case of ocular parastrongyliasis from Sri Lanka. *Trans. R. Soc. Trop. Med. Hyg.* **1998**, *92*, 64–65. [CrossRef] [PubMed]
74. Alibhoy, A.T.; Senanayake, B.; Fernando, M.A.; Amarasekera, H.S.; Wijesekera, J.C. A case of eosinophilic meningitis. *Ceylon Med. J.* **1999**, *44*, 173–174.

75. Dissanaike, A.S.; Ihalamulla, R.L.; Naotunne, T.S.; Senarathna, T.; Withana, D.S. Third report of ocular parastrongyliasis (angiostrongyliasis) from Sri Lanka. *Parassitologia* **2001**, *43*, 95–97.
76. Ilhalamulla, R.; Fernando, S.; Weerasena, K.; Cross, J.; Dissanaike, A.; Fonseka, C. A further case of Parastrongyliasis (=Angiostrongyliasis) from the eye of a patient in Sri Lanka. *Ceylon J. Med. Sci.* **2008**, *50*, 15. [CrossRef]
77. Paul, A.; Pammal, A.T. Ocular parasitosis: A rare cause of hypertensive uveitis. *Indian J. Ophthalmol.* **2008**, *56*, 501–502. [CrossRef]
78. Sah, R.; Khatri, A.; Kharel, R.; Kc, H.; Rabaan, A.A.; Tiwari, R.; Dhama, K.; Malik, Y.S.; Donovan, S.; Rodriguez-Morales, A.J.; et al. Case report: Management of dead intraocular helminth parasites in asymptomatic patients. *Am. J. Trop. Med. Hyg.* **2020**, *103*, 719–722. [CrossRef]
79. Peng, J.; He, Z.-P.; Zhang, S.; Lun, Z.-R.; Wu, Z.-D.; Fan, C.-K.; Brown, C.L.; Cheng, P.-C.; Peng, S.-Y.; Yang, T.-B. Phylogeography of *Angiostrongylus cantonensis* (Nematoda: Angiostrongylidae) in southern China and some surrounding areas. *PLoS Negl. Trop. Dis.* **2017**, *11*, e0005776. [CrossRef]
80. Tokiwa, T.; Hashimoto, T.; Yabe, T.; Komatsu, N.; Akao, N.; Ohta, N. First report of *Angiostrongylus cantonensis* (Nematoda: Angiostrongylidae) infections in invasive rodents from five islands of the Ogasawara Archipelago, Japan. *PLoS ONE* **2013**, *8*, e70729. [CrossRef]
81. Alicata, J.E.; McCarthy, D.D. On the incidence and distribution of the rat lungworm *Angiostrongylus cantonensis* in the cook islands, with observations made in new zealand and western samoa. *Can. J. Zool.* **1964**, *42*, 605–611. [CrossRef]
82. Kliks, M.M.; Palumbo, N.E. Eosinophilic meningitis beyond the Pacific Basin: The global dispersal of a peridomestic zoonosis caused by *Angiostrongylus cantonensis*, the nematode lungworm of rats. *Soc. Sci. Med.* **1992**, *34*, 199–212. [CrossRef]
83. Alicata, J.E. The presence of *Angiostrongylus cantonensis* in islands of the Indian ocean and probable role of the giant African snail, *achatina fulica*, in dispersal of the parasite to the pacific islands. *Can. J. Zool.* **1966**, *44*, 1041–1049. [CrossRef] [PubMed]
84. Yousif, F.; Ibrahim, A. The first record of *Angiostrongylus cantonensis* from Egypt. *Parasitol. Res.* **1978**, *56*, 73–80. [CrossRef]
85. Aguiar, P.H.; Morera, P.; Pascual, J. First record of *Angiostrongylus cantonensis* in Cuba. *Am. J. Trop. Med. Hyg.* **1981**, *30*, 963–965. [CrossRef]
86. Vargas, M.; JD, G.P.; Malek, E.A. First record of *Angiostrongylus cantonensis* (Chen, 1935) (Nematoda: Metastrongylidae) in the Dominican Republic. *Trop. Med. Parasitol.* **1992**, *43*, 253–255. [PubMed]
87. Campbell, B.G.; Little, M.D. The finding of *Angiostrongylus cantonensis* in rats in New Orleans. *Am. J. Trop. Med. Hyg.* **1988**, *38*, 568–573. [CrossRef] [PubMed]
88. Caldeira, R.L.; Mendonça, C.L.G.F.; Goveia, C.O.; Lenzi, H.L.; Graeff-Teixeira, C.; Lima, W.S.; Mota, E.M.; Pecora, I.L.; Medeiros, A.M.Z.D.; Carvalho, O.D.S. First record of molluscs naturally infected with *Angiostrongylus cantonensis* (Chen, 1935) (Nematoda: Metastrongylidae) in Brazil. *Mem. Inst. Oswaldo Cruz* **2007**, *102*, 887–889. [CrossRef]
89. Foronda, P.; López-González, M.; Miquel, J.; Torres, J.; Segovia, M.; Abreu-Acosta, N.; Casanova, J.C.; Valladares, B.; Mas-Coma, S.; Bargues, M.D.; et al. Finding of *Parastrongylus cantonensis* (Chen, 1935) in *Rattus rattus* in Tenerife, Canary Islands (Spain). *Acta Trop.* **2010**, *114*, 123–127. [CrossRef]
90. Paredes-Esquivel, C.; Sola, J.; Delgado-Serra, S.; Riera, M.P.; Negre, N.; Miranda, M.Á.; Jurado-Rivera, J.A. *Angiostrongylus cantonensis* in North African hedgehogs as vertebrate hosts, Mallorca, Spain, October 2018. *Eurosurveillance* **2019**, *24*, 1900489. [CrossRef]
91. Mugisha, L.; Bwangamoi, O.; Cranfield, M. *Angiostrongylus cantonensis* and other parasites infections of rodents of Budongo Forest Reserve, Uganda. *Afr. J. Anim. Biomed. Sci.* **2012**, *7*, 91–95.
92. Iwanowicz, D.D.; Sanders, L.R.; Schill, W.B.; Xayavong, M.V.; da Silva, A.J.; Qvarnstrom, Y.; Smith, T. Spread of the rat lungworm (*Angiostrongylus cantonensis*) in giant African land snails (*Lissachatina fulica*) in Florida, USA. *J. Wildl. Dis.* **2015**, *51*, 749–753. [CrossRef] [PubMed]
93. Rizor, J.; Yanez, R.A.; Thaiwong, T.; Kiupel, M. *Angiostrongylus cantonensis* in a red ruffed lemur at a zoo, Louisiana, USA. *Emerg. Infect. Dis.* **2022**, *28*, 1058. [CrossRef]
94. Dissanaike, A.S.; Cross, J.H. Ocular parastrongyliasis (=angiostrongyliasis): Probable first report of human infection from a patient in Ceylon (Sri Lanka). *Parassitologia* **2004**, *46*, 315–316. [PubMed]
95. Limaye, L.S.; Pradhan, V.R.; Bhopale, M.K.; Renapurkar, D.M.; Sharma, K.D. *Angiostrongylus cantonensis* study of intermediate paratenic and definitive hosts in greater Bombay India. *Helminthologia* **1988**, *25*, 31–35.
96. Limaye, L.S.; Bhopale, M.K.; Renapurkar, D.M.; Sharma, K.D. The distribution of *Angiostrongylus cantonensis* (Chen) in the central nervous system of laboratory rats. *Folia Parasitol.* **1983**, *30*, 281–284.
97. Mahajan, R.; Almeida, A.; Sengupta, S.; Renapurkar, D. Seasonal intensity of *Angiostrongylus cantonensis* in the intermediate host, *Laevicaulis alte*. *Int. J. Parasitol.* **1992**, *22*, 669–671. [CrossRef] [PubMed]
98. York, E.M.; Creecy, J.P.; Lord, W.D.; Caire, W. Geographic range expansion for rat lungworm in North America. *Emerg. Infect. Dis.* **2015**, *21*, 1234–1236. [CrossRef]
99. Agrawal, V.C. *Taxonomic Studies on Indian Muridae and Hystricidae (Mammalia: Rodentia)*; Rec. zool. Surv. India, Occasional Paper No. 180. i-viii; Director, Zoological Survey of India: Calcutta, India, 2000; pp. 1–186.
100. Wilson, D.E.; Reeder, D.M. *Mammal Species of the World: A Taxonomic and Geographic Reference*, 3rd ed.; JHU Press: Baltimore, MA, USA, 2005; ISBN 0-8018-8221-4.
101. Prakash, I.; Singh, P.; Nameer, P.; Ramesh, D.; Molur, S. South asian muroids. *Mamm. S. Asia* **2015**, *2*, 574–642.

102. Wilson, D.E.; Lacher, T.E.; Mittermeier, R.A. *Handbook of the Mammals of the World*; Rodents II; Lynx Edicions: Barcelona, Spain, 2017; Volume 7.
103. Dissanaike, A.S. *The Proper Study of Mankind*; Ceylon Association for the Advancement of Science: Colombo, Sri Lanka, 1968; pp. 115–142.
104. Yong, H.S.; Eamsobhana, P. Definitive rodent hosts of the rat lungworm *Angiostrongylus cantonensis*. *Raffles Bull. Zool.* **2013**, *29*, 111–115.
105. Adhikari, P.; Han, S.-H.; Kim, Y.-K.; Kim, T.-W.; Thapa, T.B.; Subedi, N.; Kunwar, A.; Banjade, M.; Oh, H.-S. New record of the Oriental house rat, *Rattus tanezumi*, in Nepal inferred from mitochondrial cytochrome B gene sequences. *Mitochondrial DNA Part B* **2018**, *3*, 386–390. [CrossRef]
106. Aplin, K.P.; Brown, P.R.; Jacob, J.; Krebs, C.J.; Singleton, G.R. *Field Methods for Rodent Studies in Asia and the Indo-Pacific*; Australian Centre for International Agricultural Research: Melbourne Australia, 2003; ISBN 1863203931.
107. Bentley, E.; Bathard, A. The rats of Addu Atoll, Maldive Islands. *Ann. Mag. Nat. Hist.* **1959**, *2*, 365–368. [CrossRef]
108. Cowie, R.H. Biology, systematics, life cycle, and distribution of *Angiostrongylus cantonensis*, the cause of rat lungworm disease. *Hawaii J. Med. Public Health* **2013**, *72*, 6. [PubMed]
109. Lv, S.; Zhang, Y.; Liu, H.-X.; Hu, L.; Yang, K.; Steinmann, P.; Chen, Z.; Wang, L.-Y.; Utzinger, J.; Zhou, X.-N. Invasive snails and an emerging infectious disease: Results from the first national survey on *Angiostrongylus cantonensis* in China. *PLoS Negl. Trop. Dis.* **2009**, *3*, e368. [CrossRef] [PubMed]
110. Richards, C.S.; Merritt, J.W. Studies on *Angiostrongylus cantonensis* in molluscan intermediate hosts. *J. Parasitol.* **1967**, *53*, 382. [CrossRef] [PubMed]
111. Tripathy, B.; Mukhopadhayay, A. Freshwater molluscs of India: An insight of into their diversity, distribution and conservation. In *Aquatic Ecosystem: Biodiversity, Ecology and Conservation*; Springer: New Delhi, India, 2015; pp. 163–195. [CrossRef]
112. Sen, S.; Ravikanth, G.; Aravind, N. Land snails (Mollusca: Gastropoda) of India: Status, threats and conservation strategies. *J. Threat. Taxa* **2012**, *4*, 3029–3037. [CrossRef]
113. Boominathan, M.; Ravikumar, G.; Chandran, M.D.S.; Ramachandra, T.V. Mangrove associated molluscs of India. In Proceedings of the National Conference on Conservation and Management of Wetland Ecosystems, Kerala, India, 6–9 November 2012; Volume 7, pp. 1–11.
114. Ratnapala, R. Land snails: Distribution and notes on ecology. In *Ecology and Biogeography in Sri Lanka*; Springer: Dordrecht, The Netherlands, 1984; pp. 391–411. [CrossRef]
115. Strong, E.E.; Gargominy, O.; Ponder, W.; Bouchet, P. Global diversity of gastropods (Gastropoda; Mollusca) in freshwater. *Freshw. Anim. Divers. Assess.* **2008**, *198*, 149–166. [CrossRef]
116. Mavinkurve, R.G.; Shanbhag, S.P.; Madhyastha, N. Non-marine molluscs of Western Ghats: A status review. *Zoos' Print J.* **2004**, *19*, 1708–1711. [CrossRef]
117. Thiengo, S.; Maldonado, A.; Mota, E.; Torres, E.; Caldeira, R.; Carvalho, O.; Oliveira, A.; Simões, R.; Fernandez, M.; Lanfredi, R. The giant African snail *Achatina fulica* as natural intermediate host of *Angiostrongylus cantonensis* in Pernambuco, northeast Brazil. *Acta Trop.* **2010**, *115*, 194–199. [CrossRef]
118. Wang, Q.-P.; Chen, X.-G.; Lun, Z.-R. Invasive fresh water snail, China. *Emerg. Infect. Dis.* **2007**, *13*, 1119–1120. [CrossRef]
119. Cowie, R.H. *Angiostrongylus cantonensis*: Agent of a sometimes fatal globally emerging infectious disease (rat lungworm disease). *ACS Chem. Neurosci.* **2017**, *8*, 2102–2104. [CrossRef] [PubMed]
120. Eamsobhana, P.; Yong, H.S.; Prasartvit, A.; Wanachiwanawin, D.; Boonyong, S. Geographical distribution and incidence of Angiostrongylus lungworms (Nematoda: Angiostrongylidae) and their rodent hosts in Thailand. *Trop. Biomed.* **2016**, *33*, 35–44. [PubMed]
121. Hamilton, L.J.; Tagami, Y.; Kaluna, L.; Jacob, J.; Jarvi, S.I.; Follett, P. Demographics of the semi-slug *Parmarion martensi*, an intermediate host for *Angiostrongylus cantonensis* in Hawai'i, during laboratory rearing. *Parasitology* **2021**, *148*, 153–158. [CrossRef] [PubMed]
122. Kim, J.R.; Wong, T.M.; Curry, P.A.; Yeung, N.W.; Hayes, K.A.; Cowie, R.H. Modelling the distribution in Hawaii of *Angiostrongylus cantonensis* (rat lungworm) in its gastropod hosts. *Parasitology* **2019**, *146*, 42–49. [CrossRef] [PubMed]
123. Tesana, S.; Srisawangwong, T.; Sithithaworn, P.; Laha, T.; Andrews, R. Prevalence and intensity of infection with third stage larvae of *Angiostrongylus cantonensis* in mollusks from Northeast Thailand. *Am. J. Trop. Med. Hyg.* **2009**, *80*, 983–987. [CrossRef]
124. Xu, Y.; Wang, W.; Yao, J.; Yang, M.; Guo, Y.; Deng, Z.; Mao, Q.; Li, S.; Duan, L. Comparative proteomics suggests the mode of action of a novel molluscicide against the invasive apple snail *Pomacea canaliculata*, intermediate host of *Angiostrongylus cantonensis*. *Mol. Biochem. Parasitol.* **2022**, *247*, 111431. [CrossRef]
125. Yang, T.-B.; Wu, Z.-D.; Lun, Z.-R. The apple snail *Pomacea canaliculata*, a novel vector of the rat lungworm, *Angiostrongylus cantonensis*: Its introduction, spread, and control in China. *Hawaii J. Med. Public Health* **2013**, *72*, 23.
126. Baloch, W.A.; Memon, U.N.; Burdi, G.H.; Soomro, A.N.; Tunio, G.R.; Khatian, A.A. Invasion of channelled apple snail *Pomacea canaliculata*, Lamarck (Gastropoda: Ampullariidae) in Haleji Lake, Pakistan. *Sindh Univ. Res. J. SURJ (Sci. Ser.)* **2012**, *44*, 263–266.
127. Joshi, R.C. Problems with the management of the golden apple snail Pomacea canaliculata: An important exotic pest of rice in Asia. In *Proceedings of the Area-Wide Control of Insect Pests: From Research to Field Implementation*; Springer: Berlin/Heidelberg, Germany, 2007; pp. 257–264.

128. Rekha Sarma, R.; Munsi, M.; Neelavara Ananthram, A. Effect of climate change on invasion risk of giant African snail (*Achatina fulica* Férussac, 1821: Achatinidae) in India. *PLoS ONE* **2015**, *10*, e0143724. [CrossRef]
129. Saha, C.; Pramanik, S.; Chakraborty, J.; Parveen, S.; Aditya, G. Abundance and body size of the invasive snail *Physa acuta* occurring in Burdwan, West Bengal, India. *J. Entomol. Zool. Stud.* **2016**, *4*, 490–497.
130. Saha, C.; Parveen, S.; Chakraborty, J.; Pramanik, S.; Aditya, G. Life table estimates of the invasive snail *Physa acuta* Draparnaud, 1805, occurring in India. *Ekologia* **2017**, *36*, 60–68. [CrossRef]
131. Paul, P.; Aditya, G. Invasion of the freshwater snail *Physella acuta* (Draparnaud, 1805) in selected ponds of North Dinajpur, India. *J. Environ. Biol.* **2021**, *42*, 577–581. [CrossRef]
132. Raut, S.K.; Mandal, R.N. Natural history of the garden slug *Laevicaulis alte*. *J. Bengal Nat. Hist. Soc.* **1984**, *3*, 104–105.
133. Thakuri, B.; Acharya, B.K.; Sharma, G. Population density and damage of invasive giant African snail *Achatina fulica* in organic farm in east Sikkim, India. *Indian J. Ecol.* **2019**, *46*, 631–635.
134. Sridhar, V.; Vinesh, L.S.; Jayashankar, M. Mapping the potential distribution of *Achatina fulica* (Bowdich) (Stylommatophora: Achatinidae) in India using CLIMEX, a bioclimatic software. *Pest Manag. Hortic. Ecosyst.* **2014**, *20*, 14–21.
135. Husain, A.; Husain, H.J.; Hasan, W.; Kendra, J.K.V. New records of tropical leather-leaf slug *Laevicaulis alte* (Ferussac, 1822) from Dehra Dun (Uttarakhand) and Jamshedpur (Jharkhand), India. *Int. J. Agric. Appl. Sci.* **2021**, *2*, 145–150. [CrossRef]
136. Baghele, M.; Mishra, S.; Meyer-Rochow, V.B.; Jung, C.; Ghosh, S. Utilization of snails as food and therapeutic agents by Baiga tribals of Baihar tehsil of Balaghat District, Madhya Pradesh, India. *Int. J. Indust. Entomol.* **2021**, *43*, 78–84.
137. Dhiman, V.; Pant, D. Human health and snails. *J Immunoassay Immunochem.* **2021**, *42*, 211–235. [CrossRef] [PubMed]
138. Prabhakar, A.K.; Roy, S. Ethno-medicinal uses of some shell fishes by people of Kosi river basin of North-Bihar, India. *Stud. Ethno-Med.* **2009**, *3*, 1–4. [CrossRef]
139. Sarkar, T.; Debnath, S.; Das, B.K.; Das, M. Edible fresh water molluscs diversity in the different water bodies of Gangarampur Block, Dakshin Dinajpur, West Bengal. *Eco. Env. Cons.* **2021**, *27*, S293–S296.
140. Baby, R.L.; Hasan, I.; Kabir, K.A.; Naser, M.N. Nutrient analysis of some commercially important molluscs of Bangladesh. *J. Sci. Res.* **2010**, *2*, 390–396. [CrossRef]
141. Rabha, H.P.; Mazumdar, M.; Baruah, U.K. Indigenous technical knowledge on ethnic dishes of snail in Goalpara district of India. *Sch. Acad. J. Biosci.* **2014**, *2*, 307–317.
142. Borkakati, R.N.; Gogoi, R.; Borah, B.K. Snail: From present perspective to the history of Assam. *Asian Agrihist.* **2009**, *13*, 227–234.
143. Ghosh, S.; Jung, C.; Meyer-Rochow, V.B. Snail farming: An Indian perspective of a potential tool for food security. *Ann. Aquac. Res.* **2016**, *3*, 1024.
144. Bagde, N.; Jain, S. Study of traditional man-animal relationship in Chhindwara district of Madhya Pradesh, India. *J. Glob. Biosci.* **2015**, *4*, 1456–1463.
145. Tsai, H.-C.; Lin, H.-H.; Wann, S.-R.; Liu, Y.-C.; Huang, C.-K.; Lai, P.-H.; Yen, M.-Y.; Kunin, C.M.; Ger, L.-P.; Lee, S.S.-J.; et al. Eosinophilic meningitis caused by *Angiostrongylus cantonensis* associated with eating raw snails: Correlation of brain magneticresonance imaging scans with clinical findings. *Am. J. Trop. Med. Hyg.* **2003**, *68*, 281–285. [CrossRef]
146. Alicata, J.E. Effect of freezing and boiling on the infectivity of third-stage larvae of *Angiostrongylus cantonensis* present in and snails and freshwater prawns. *J. Parasitol.* **1967**, *53*, 1064. [CrossRef] [PubMed]
147. Crook, J.R.; Fulton, S.E.; Supanwong, K. The infectivity of third stage *Angiostrongylus cantonensis* larvae shed from drowned *Achatina fulica* snails and the effect of chemical agents on infectivity. *Trans. R. Soc. Trop. Med. Hyg.* **1971**, *65*, 602–605. [CrossRef] [PubMed]
148. Prasad, S.; Kumar, M.U.; Kumari, A. Freshwater shellfish, *Pila globosa* (Gastropoda) favourable endeavour for rural nutrition. *Agric. Lett.* **2021**, *2*, 32–35.
149. Radomyos, P.; Tungtrongchitr, A.; Praewanich, R. Experimental infection of yellow tree monitor (*Varanus bengalensis*) and related species with *Angiostrongylus cantonensis*. *Southeast Asian J. Trop. Med. Public Health* **1992**, *23*, 167–168. [PubMed]
150. Böhme, W. Checklist of the living monitor lizards of the world (Family Varanidae). *Zool. Verh.* **2003**, *341*, 4–43.
151. Pianka, E.; King, D. *Varanoid Lizards of the World*; Indiana University Press: Bloomington, IN, USA, 2004; ISBN 0253013690.
152. Aengals, R.; Kumar, V.S.; Palot, M.J.; Ganesh, S.R. *A Checklist of Reptiles of India*; Zoological Survey of India: Kolkata, India, 2018; p. 35.
153. Auffenberg, W.; Rahman, H.; Iffat, F.; Perveen, Z. A study of *Varanus flavescens* (Hardwicke & Gray) (Sauria: Varanidae). *J. Bombay Nat. Hist. Soc.* **1989**, *86*, 286–307.
154. Wikramanayake, E.D.; Dryden, G.L. Thermal ecology of habitat and microhabitat use by sympatric *Varanus bengalensis* and *V. salvator* in Sri Lanka. *Copeia* **1993**, 709–714. [CrossRef]
155. Radomyos, P.; Tungtrongchitr, A.; Praewanich, R.; Khewwatchan, P.; Kantangkul, T.; Junlananto, P.; Ayudhya, S.I. Occurrence of the infective stage of *Angiostrongylus cantonensis* in the yellow tree monitor (*Varanus bengalensis*) in five Provinces of Thailand. *Southeast Asian J. Trop. Med. Public Health* **1994**, *25*, 498–500.
156. Anettová, L.; Izquierdo-Rodriguez, E.; Foronda, P.; Baláž, V.; Novotný, L.; Modrý, D. Endemic lizard *Gallotia galloti* is a paratenic host of invasive *Angiostrongylus cantonensis* in Tenerife, Spain. *Parasitology* **2022**, *149*, 934–939. [CrossRef] [PubMed]
157. Yang, L.; Darasavath, C.; Chang, K.; Vilay, V.; Sengduangphachanh, A.; Adsamouth, A.; Vongsouvath, M.; Keolouangkhot, V.; Robinson, M.T. Cluster of angiostrongyliasis cases following consumption of raw monitor lizard in the Lao People's Democratic Republic and review of the literature. *Trop. Med. Infect. Dis.* **2021**, *6*, 107. [CrossRef] [PubMed]

158. Ash, L.R. The Occurrence of *Angiostrongylus cantonensis* in frogs of New Caledonia with observations on paratenic hosts of metastrongyles. *J. Parasitol.* **1968**, *54*, 432. [CrossRef]
159. Chellappandian, M.; Pandikumar, P.; Mutheeswaran, S.; Paulraj, M.G.; Prabakaran, S.; Duraipandiyan, V.; Ignacimuthu, S.; Al-Dhabi, N. Documentation and quantitative analysis of local ethnozoological knowledge among traditional healers of Theni district, Tamil Nadu, India. *J. Ethnopharmacol.* **2014**, *154*, 116–130. [CrossRef]
160. Jugli, S.; Chakravorty, J.; Meyer-Rochow, V.B. Zootherapeutic uses of animals and their parts: An important element of the traditional knowledge of the Tangsa and Wancho of Eastern Arunachal Pradesh, North-East India. *Environ. Dev. Sustain.* **2020**, *22*, 4699–4734. [CrossRef]
161. Chakravorty, J.; Meyer-Rochow, V.B.; Ghosh, S. Vertebrates used for medicinal purposes by members of the Nyishi and Galo tribes in Arunachal Pradesh (North-East India). *J. Ethnobiol. Ethnomed.* **2011**, *7*, 13. [CrossRef]
162. Das, K.S.; Choudhury, S.; Nonglait, K.C.L. Zootherapy among the ethnic groups of north eastern region of india-a critical review. *J. Crit. Rev.* **2017**, *4*, 1–9. [CrossRef]
163. Mahawar, M.M.; Jaroli, D.P. Animals and their products utilized as medicines by the inhabitants surrounding the Ranthambhore National Park, India. *J. Ethnobiol. Ethnomed.* **2006**, *2*, 46. [CrossRef]
164. Verma, A.K.; Prasad, S.B.; Rongpi, T.; Arjun, J. Traditional healing with animals (zootherapy) by the major ethnic group of Karbi Anglong district of Assam, India. *Int. J. Pharm. Pharm. Sci.* **2014**, *6*, 593–600.
165. Borah, M.P.; Prasad, S.B. Ethnozoological study of animals based medicine used by traditional healers and indigenous inhabitants in the adjoining areas of Gibbon Wildlife Sanctuary, Assam, India. *J. Ethnobiol. Ethnomed.* **2017**, *13*, 39. [CrossRef]
166. Betlu, A.L.S. Indigenous knowledge of zootherapeutic use among the Biate tribe of Dima Hasao District, Assam, Northeastern India. *J. Ethnobiol. Ethnomed.* **2013**, *9*, 56. [CrossRef] [PubMed]
167. Oku, Y.; Katakura, K.; Kamiya, M. Tadpole of the clawed frog, *Xenopus Laevis*, as an experimental intermediate host of *Angiostrongylus cantonensis*. *Am. J. Trop. Med. Hyg.* **1980**, *29*, 316–318. [CrossRef] [PubMed]
168. Niebuhr, C.N.; Jarvi, S.I.; Kaluna, L.; Fischer, B.L.T.; Deane, A.R.; Leinbach, I.L.; Siers, S.R. Occurrence of rat lungworm (*Angiostrongylus cantonensis*) in invasive coqui frogs (*Eleutherodactylus coqui*) and other hosts in Hawaii, USA. *J. Wildl. Dis.* **2020**, *56*, 203. [CrossRef] [PubMed]
169. Asato, R.; Sato, Y.; Otsuru, M. The occurrence of *Angiostrongylus cantonensis* in toads and frogs in Okinawa Prefecture, Japan. *Jpn J. Parasitol.* **1978**, *27*, 1–8.
170. Mahawar, M.M.; Jaroli, D. Traditional zootherapeutic studies in India: A review. *J. Ethnobiol. Ethnomed.* **2008**, *4*, 17. [CrossRef]
171. Shah, K.B.; Gire, M.K. Some Amphibians and Their Local Uses in Arun Basin. *J. Nat. Hist. Museum.* **1992**, *13*, 9–17.
172. Pradhan, S.; Mishra, D.; Sahu, K.R. Herpetofauna used as traditional medicine by tribes of Gandhamardan Hills Range, Western Orissa, India. *Int. J. Res. Zool.* **2014**, *4*, 32–35.
173. Pandian, T.J.; Marian, M.P. Production and utilization of frogs: An ecological view. *Proc. Anim. Sci.* **1986**, *95*, 289–301. [CrossRef]
174. Kawlni, L.; Lalronunga, S.; Manral, U.; Das, A. Uchang-Sa: A report on frog consumption around Dampa Tiger Reserve, Mizoram, Northeast India. *Hamadryad* **2022**, *39*, 30–36.
175. Talukdar, S.; Sengupta, S. Edible frog species of Nagaland. *J. Environ. Biol.* **2020**, *41*, 927–930. [CrossRef]
176. Talukdar, S.; Sengupta, S.; Konyak, M.; Rizwan, M. Cooking methods and medicinal uses of frog species among the Naga tribes in Dimapur. *World News Nat. Sci.* **2020**, *30*, 144–149.
177. Thomas, A.; Biju, S.D. Tadpole consumption is a direct threat to the endangered purple frog, *Nasikabatrachus Sahyadrensis*. *Salamandra* **2015**, *51*, 252–258.
178. Yong, H.-S.; Eamsobhana, P.; Song, S.-L.; Prasartvit, A.; Lim, P.-E. Molecular phylogeography of *Angiostrongylus cantonensis* (Nematoda: Angiostrongylidae) and genetic relationships with congeners using cytochrome b gene marker. *Acta Trop.* **2015**, *148*, 66–71. [CrossRef] [PubMed]
179. Yong, H.-S.; Song, S.-L.; Eamsobhana, P.; Goh, S.-Y.; Lim, P.-E. Complete mitochondrial genome reveals genetic diversity of *Angiostrongylus cantonensis* (Nematoda: Angiostrongylidae). *Acta Trop.* **2015**, *152*, 157–164. [CrossRef]
180. Vitta, A.; Srisongcram, N.; Thiproaj, J.; Wongma, A.; Polsut, W.; Fukruksa, C.; Yimthin, T.; Mangkit, B.; Thanwisai, A.; Dekumyoy, P. Phylogeny of *Angiostrongylus cantonensis* in Thailand based on cytochrome c oxidase subunit I gene sequence. *Southeast Asian J. Trop. Med. Public Health* **2016**, *47*, 377–386.
181. Dumidae, A.; Subkrasae, C.; Ardpairin, J.; Thanwisai, A.; Vitta, A. Genetic analysis of a 66-kDa protein-encoding gene of *Angiostrongylus cantonensis* and *Angiostrongylus malaysiensis*. *Parasitology* **2023**, *150*, 98–114. [CrossRef]
182. Rodpai, R.; Intapan, P.M.; Thanchomnang, T.; Sanpool, O.; Sadaow, L.; Laymanivong, S.; Aung, W.P.; Phosuk, I.; Laummaunwai, P.; Maleewong, W. *Angiostrongylus cantonensis* and *A. malaysiensis* broadly overlap in Thailand, Lao PDR, Cambodia and Myanmar: A molecular survey of larvae in land snails. *PLoS ONE* **2016**, *11*, e0161128. [CrossRef] [PubMed]
183. Chan, A.H.E.; Chaisiri, K.; Dusitsittipon, S.; Jakkul, W.; Charoennitiwat, V.; Komalamisra, C.; Thaenkham, U. Mitochondrial ribosomal genes as novel genetic markers for discrimination of closely related species in the *Angiostrongylus cantonensis* lineage. *Acta Trop.* **2020**, *211*, 105645. [CrossRef] [PubMed]
184. Eamsobhana, P.; Lim, P.; Yong, H. Phylogenetics and systematics of *Angiostrongylus* lungworms and related taxa (Nematoda: Metastrongyloidea) inferred from the nuclear small subunit (SSU) ribosomal DNA sequences. *J. Helminthol.* **2014**, *89*, 317–325. [CrossRef]

185. Eamsobhana, P.; Song, S.-L.; Yong, H.-S.; Prasartvit, A.; Boonyong, S.; Tungtrongchitr, A. Cytochrome c oxidase subunit I haplotype diversity of *Angiostrongylus cantonensis* (Nematoda: Angiostrongylidae). *Acta Trop.* **2017**, *171*, 141–145. [CrossRef]
186. Eamsobhana, P.; Yong, H.-S.; Song, S.-L.; Gan, X.-X.; Prasartvit, A.; Tungtrongchitr, A. Molecular phylogeography and genetic diversity of *Angiostrongylus cantonensis* and *A. malaysiensis* (Nematoda: Angiostrongylidae) based on 66-kDa protein gene. *Parasitol. Int.* **2019**, *68*, 24–30. [CrossRef]
187. Mishra, N.; Panda, T. Zootherapeutical Uses of Animal Diversity in Coastal District of Orissa, India. *Br. J. Pharmacol. Toxicol.* **2011**, *2*, 154–158.
188. Joshi, N.; Pandey, S. Meat Demand-Snailed It: A Comprehensive Review on Snail Rearing, to Meet the Meat Demand in Future India. *J. Entomol. Zool. Stud* **2019**, *7*, 396–400.
189. MESHRAM, L.; HUSAIN, N. Use of Animals as Traditional Medicine in Chhattisgarh, India. *Indian J. Sci. Res* **2017**, *12*, 26–29.
190. Raja, L.; Matheswaran, P.; Anbalagan, M.; Sureshkumar, V.; Ganesan, D.; Gani, S.B. Ethnozoological Study of Animal-Based Products Practices among the Tribal Inhabitants in Kolli Hills Namakkal District, Tamil Nadu, India. *World J. Pharm. Pharm. Sci.* **2018**, *7*, 785–797.
191. Debnath, C.; Sahoo, L.; Haldar, A.; Datta, M.; Yadav, G.S.; Singha, A.; Bhattacharjee, J. Proximate and Mineral Composition of Freshwater Snails of Tripura, North-East India. *Fish. Technol.* **2016**, *53*, 307–312.
192. Chinlampianga, M.; Singh, R.K.; Shukla, A.C. *Ethnozoological Diversity of Northeast India: Empirical Learning with Traditional Knowledge Holders of Mizoram and Arunachal Pradesh*; NISCAIR-CSIR: Delhi, India, 2013.
193. Singh, R.K.; Kumar, A.; Singh, A.; Singhal, P. Evidence That Cultural Food Practices of Adi Women in Arunachal Pradesh, India, Improve Social-Ecological Resilience: Insights for Sustainable Development Goals. *Ecol. Process.* **2020**, *9*, 1–19. [CrossRef]
194. Jadhav, A.; Das, N.K.; Aravind, N.A. Edible Freshwater Molluscs from Northeast India. *Tentacle* **2020**, *28*, 1–4.
195. Bhattacharya, S.; Chakraborty, M.; Bose, M.; Mukherjee, D.; Roychoudhury, A.; Dhar, P.; Mishra, R. Indian Freshwater Edible Snail *Bellamya Bengalensis* Lipid Extract Prevents T Cell Mediated Hypersensitivity and Inhibits LPS Induced Macrophage Activation. *J. Ethnopharmacol.* **2014**, *157*, 320–329. [CrossRef] [PubMed]
196. Gomes, A.; Sarkar, A.; Ghosh, S.; Saha, K.; Saha, P.P. Edible Fresh Water Snail *Viviparous Bengalensis* Purified Flesh Protein VB-P4 Induced Toxicities and Its Protection by Heat Treatment. *J. Toxins* **2015**, *2*, 3.
197. Dhakal, P.; Chettri, B.; Lepcha, S.; Acharya, B.K. Rich yet Undocumented Ethnozoological Practices of Socio-Culturally Diverse Indigenous Communities of Sikkim Himalaya, India. *J. Ethnopharmacol.* **2020**, *249*, 112386. [CrossRef]
198. Patel, R.; Pal, M.C. Exploring Ethnozoological Observations Among The Tribal Inhabitants In 'Bajag' Forest Range Of Dindori District Of Central India. *Int. J. Aquat. Sci.* **2021**, *12*, 4513–4538.
199. Vyas, N.; Mahawar, M.M.; Jaroli, D.P. Traditional Medicines Derived from Domestic Animals Used by Rebari Community of Rajasthan, India. *Nature* **1970**, *7*, 129–138. [CrossRef]
200. Solavan, A.; Paulmurugan, R.; Wilsanand, V.; Sing, A.J. *Traditional Therapeutic Uses of Animals among Tribal Population of Tamil Nadu*; CSIR: Lucknow, India, 2004.
201. Haq, S.M.; Calixto, E.S.; Yaqoob, U.; Ahmed, R.; Mahmoud, A.H.; Bussmann, R.W.; Mohammed, O.B.; Ahmad, K.; Abbasi, A.M. Traditional Usage of Wild Fauna among the Local Inhabitants of Ladakh, Trans-Himalayan Region. *Animals* **2020**, *10*, 2317. [CrossRef]
202. Imkongwapang, R.; Jyrwa, D.B.; Lal, P.; Tandon, V. A Checklist of Helminth Parasite Fauna in Anuran Amphibia (Frogs) of Nagaland, Northeast India. *J. Parasit. Dis.* **2014**, *38*, 85–100. [CrossRef] [PubMed]
203. Hassan, M.; Haq, S.M.; Amjad, M.S.; Ahmad, R.; Bussmann, R.W.; de la Lastra, J.M.P. Invertebrates and Herptiles for Livelihoods—Ethnozoological Use among Different Ethnic Communities in Jammu and Kashmir (Indian Himalayas). *Front. Pharmacol.* **2023**, *13*, 1043155. [CrossRef] [PubMed]
204. Narzary, J.; Bordoloi, S. Ethnozoological Practices on Frogs of Bodo Tribe from Kokrajhar District, Assam, India. *Am. J. Ethnomedicine* **2014**, *1*, 368–370.
205. Vijayakumar, S.; Yabesh, J.E.M.; Prabhu, S.; Ayyanar, M.; Damodaran, R. Ethnozoological Study of Animals Used by Traditional Healers in Silent Valley of Kerala, India. *J. Ethnopharmacol.* **2015**, *162*, 296–305. [CrossRef] [PubMed]
206. Joseph, J.; Sreeedharan, S.; George, S.; Antony, M.M. The Complete Mitochondrial Genome of an Endemic Cichlid Etroplus Canarensis from Western Ghats, India (Perciformes: Cichlidae) and Molecular Phylogenetic Analysis. *Mol. Biol. Rep.* **2022**, *49*, 3033–3044. [CrossRef]
207. Esakkimuthu, S.; Sylvester Darvin, S.; Mutheeswaran, S.; Gabriel Paulraj, M.; Pandikumar, P.; Ignacimuthu, S.; Al-Dhabi, N.A. A Study on Food-Medicine Continuum among the Non-Institutionally Trained Siddha Practitioners of Tiruvallur District, Tamil Nadu, India. *J. Ethnobiol. Ethnomed.* **2018**, *14*, 45. [CrossRef] [PubMed]
208. Kumar Kiran, S.; Anoop, V.S.; Sivakumar, K.C.; Dinesh, R.; Mano, J.P.; Kaushik, D.; Sanil, G. An Additional Record of Fejervarya Manoharani Garg and Biju from the Western Ghats with a Description of Its Complete Mitochondrial Genome. *Zootaxa* **2017**, *4277*, 491–502. [CrossRef]
209. Anoop, V.S.; George, S. Population Genetic Structure and Evolutionary Demographic Patterns of Phrynoderma Karaavali, an Edible Frog Species of Kerala, India. *J. Genet.* **2023**, *102*, 8. [CrossRef]

210. Chettri, B.; Bhupathy, S.; Acharya, B.K. An Overview of the Herpetofauna of Sikkim with Emphasis on the Elevational Distribution Pattern and Threats and Conservation Issues. In *Biodiversity of Sikkim: Exploring and Conserving a Global Hotspot*; Information and Public Relations Department, Government of Sikkim: Gangtok, India, 2011; pp. 233–254.
211. Jaroli, D.P.; Mahawar, M.M.; Vyas, N. An Ethnozoological Study in the Adjoining Areas of Mount Abu Wildlife Sanctuary, India. *J. Ethnobiol. Ethnomed.* **2010**, *6*, 6. [CrossRef]

pathogens

MDPI

Article

The Zoonotic *Angiostrongylus cantonensis* and the Veterinary Parasite *Aelurostrongylus abstrusus* Infecting Terrestrial Gastropods from Urban Areas of Macapá, Brazilian Amazon Region

Tatiane Alves Barbosa [1,2,3], Silvana Carvalho Thiengo [1], Monica Ammon Fernandez [1], Jucicleide Ramos-de-Souza [1,*] and Suzete Rodrigues Gomes [1]

[1] Laboratório de Malacologia, Instituto Oswaldo Cruz, Fundação Oswaldo Cruz, Av. Brasil, 4365, Rio de Janeiro 21040-900, Brazil; endemias.dvsmcp@gmail.com (T.A.B.); sthiengo@ioc.fiocruz.br (S.C.T.); ammon@ioc.fiocruz.br (M.A.F.); suzete.gomes@ioc.fiocruz.br (S.R.G.)

[2] Programa de Pós-Graduação Stricto Sensu em Vigilância e Controle de Vetores—PPG-VCV, IOC/Fiocruz, Av. Brasil 4365, Rio de Janeiro 21040-900, Brazil

[3] Departamento de Vigilância Ambiental, Divisão de Vigilância e Controle de Vetores, Secretaria Municipal de Vigilância em Saúde de Macapá, Amapá 68906-849, Brazil

* Correspondence: ramosdesouzza@gmail.com

Abstract: Metastrongyloidea includes nematodes that parasitize mammals, mainly infecting their respiratory and cardiovascular systems, and are responsible for emerging zoonosis in the world. Terrestrial mollusks are their main intermediate hosts, with few exceptions. Here we present the results of a malacological survey to know the distribution of *Angiostrongylus cantonensis* in Macapá, Amapá, in the Brazilian Amazon region, after the report of a case of eosinophilic meningitis in 2018. Mollusks were collected in 45 neighborhoods between March 2019 and February 2020. They were identified, parasitologically analyzed, and their nematodes parasites were identified based on the morphology and MT-CO1 sequencing. Infections of An. cantonensis were observed in *Achatina fulica*, *Sarasinula linguaeformis* and *Subulina octona*. These are the first records of the natural infection of the last two species by *An. cantonensis* in the Brazilian Amazon region. The angiostrongylid *Aelurostrongylus abstrusus*, which parasitizes cats, was also detected parasitizing *A. fulica* and *Diplosolenodes occidentalis*. This is also the first record of the slug *D. occidentalis* infected by *Ae. abstrusus*. The highest infection rates were recorded in neighborhoods where the environment conditions favor the proliferation of both mollusks and rodents. The results demonstrate the ample distribution of *An. cantonensis* in Macapá and the need for surveillance and mollusk vector control in Brazil and other countries.

Keywords: *Achatina fulica*; slugs; snails; zoonotic helminths; eosinophilic meningitis; *Aelurostrongylus abstrusus*

check for updates

Citation: Barbosa, T.A.; Thiengo, S.C.; Fernandez, M.A.; Ramos-de-Souza, J.; Gomes, S.R. The Zoonotic *Angiostrongylus cantonensis* and the Veterinary Parasite *Aelurostrongylus abstrusus* Infecting Terrestrial Gastropods from Urban Areas of Macapá, Brazilian Amazon Region. *Pathogens* **2024**, *13*, 255. https://doi.org/10.3390/pathogens13030255

Academic Editor: Anastasia Diakou

Received: 31 January 2024
Revised: 26 February 2024
Accepted: 13 March 2024
Published: 15 March 2024

1. Introduction

The superfamily Metastrongyloidea includes nematode species that parasitize primarily mammals, by infecting their respiratory and cardiovascular systems, although a few species are neurotropic [1]. Most metastrongyloid species, specifically from the family Angiostrongylidae, use terrestrial gastropods as intermediate hosts, although some are known to parasitize freshwater gastropods [2,3]. In addition to wild animals, some angiostrongylid species parasitize humans, and domestic and synanthropic animals. The most important of these angiostrongylid in public health terms is *Angiostrongylus cantonensis* (Chen, 1935), which causes cerebral angiostrongyliasis, which is also known as Eosinophilic Meningitis (EM). A second species of the same genus present in Brazil, *Angiostrongylus costaricensis* Morera and Céspedes, 1971, causes Abdominal Angiostrongyliasis (AA) [4–7].

While EM is naturally endemic to Southeast Asia and some Pacific islands, it is currently found in many regions around the world, such as the Americas, including Brazil [8–13]. Up to now, approximately 40 cases of EM have been recorded in Brazil, with the most recent case coming from the northern state of Amapá (AP) in 2020, in which the giant African land snail, *Achatina fulica* Bowdich, 1822, was identified as the transmission agent [12,14,15].

The life cycle of *An. cantonensis* involves rats as definitive hosts and various species of gastropods as intermediate hosts. Humans are accidentally infected in most cases through the ingestion of mollusks or infected paratenic hosts, as well as the consumption of foodstuffs contaminated with infectious third stage larvae (L_3) of the nematode, which are found in the mucus of the mollusks [16–18]. In Brazil, *Rattus rattus* (Linnaeus, 1758) and *Rattus norvegicus* (Berkenhout, 1796) have been found with natural infection of *An. cantonensis* [3,19–21].

Mollusks are infected through the ingestion of first stage larvae (L_1) released in the feces of rodents or by the active penetration of the tegument of the mollusk by the larvae. Two molts occur in the mollusk tissue (L_2 and L_3). A number of different species of mollusk are known to be infected naturally by *An. cantonensis*, which contributes to the persistence of the natural cycle of this parasite, and its dispersal in the environment: *Sarasinula marginata* (Semper, 1885), *Bradybaena similaris* (Férussac, 1821), *Subulina octona* (Bruguiére, 1789), *Pomacea lineata* (Spix in Wagner, 1827) and other *Pomacea* species. [2,8,11,13,18,22].

Another genus of concern in Angiostrongylidae, is *Aelurostrongylus abstrusus* (Railliet, 1898), which parasitizes the respiratory tract of both domestic and wild felids [23–25] that are infected through the ingestion of parasitized mollusks or paratenic hosts [26,27]. The first stage larvae go up into the trachea, where they are swallowed and excreted in the feces [25,26]. The larvae penetrate through the mollusk tissue before developing to the third larval stage. In Brazil, *A. fulica* is the mollusk most frequently associated with *Ae. abstrusus*, although infection of the native slug *Latipes erinaceus* (Colosi, 1921) has been recorded in Rio de Janeiro [3,25].

Given this scenario of public health concern and the recent report of a case of EM in Macapá [14], the present study aimed to expand the investigation of the occurrence and distribution of *An. cantonensis* and other angiostrongylid nematodes in terrestrial mollusks in this municipality. Therefore, the main goal was to know the distribution of the nematodes that affect the health of both humans and animals and their intermediate hosts in Macapá, thus contributing to guiding surveillance and control of snail-borne parasitic diseases.

2. Materials and Methods

2.1. Study Design

The terrestrial mollusks were collected from 45 neighborhoods of Macapá over a 14-month period between March 2019 and February 2020. The collecting points included vacant lots, and public and residential gardens, in which large amounts of debris or decomposing organic material were present, given that these are favorable environments for the occurrence of terrestrial mollusks and rodents. The specimens were collected manually during active searches of each site, which lasted 10–30 min. Each point was georeferenced with a handheld Garmin 64s GPS, for plotting in ArcGIS 10.4.1.

In the laboratory, the mollusks were kept alive until the parasitological analysis. Specimens from each point were fixed for taxonomic identification based on the diagnosis of their morphology (shell and anatomy of reproductive system), supported by the relevant literature [28–35]. These specimens were deposited in the Mollusk Collection of the Oswaldo Cruz Institute (CMIOC/Fiocruz) in Rio de Janeiro.

2.2. Parasitological Analysis

A total of 306 mollusk specimens were analyzed individually for the presence of nematode parasites through the artificial digestion of their tissue [36,37] for releasing of the nematode larvae. These larvae were initially identified based on their morphology under a stereomicroscope (100×) and optical microscope (40×), according to Ash [38], Thiengo et al. [39], and Rodrigues et al. [25]. The larvae with diagnostic characteristics of the Metastrongyloidea were separated for the sequencing of the mitochondrial cytochrome *c* oxidase subunit I gene (COI). The larvae of *Ae. abstrusus* were identified only by their morphology, based on the presence of a rounded, button-like structure in the terminal portion of the tail.

2.3. Molecular Analysis

For each sample identified as *Angiostrongylus* based on the morphological criteria, 10 larvae were transferred to micro-centrifuge tubes and frozen at −18 °C in 30 µL PBS (Phosphate-Buffered Saline). The genomic DNA was then isolated by thermal shock using liquid nitrogen, according to the Standard Operating Procedure (SOP) used for this technique in the Brazilian National Reference Laboratory for Schistosomiasis and Malacology (LRNEM-IOC), where the MT-CO1 was amplified [40]. The Polymerase Chain Reaction (PCR) was run in a final volume of 50 µL containing 23.90 µL of ultra-pure water, 11 µL of 10% trehalose, 5.5 µL of 10× PCR reaction buffer, 4.4 µL of 2.5 mM dNTPs, 2.75 µL of 50 mM $MgCl_2$, 1.1 µL of each primer (forward and reverse) (0.2 µM of Nem_F3 and Nem_R3, modified from Prosser et al. [40], and 0.25 µL of recombinant Taq DNA polymerase (Thermo Fisher Scientific, Waltham, MA, USA). A total of 5 µL of the DNA sample was added to the mixture, to produce a final reaction volume of 55 µL. The PCR products were purified using the Illustra GFX PCR DNA and Gel Band Purification kit (GE Healthcare, Little Chalfont, UK), following the manufacturer's protocol. The purified products were sequenced bidirectionally using the BigDye Terminator v3.1 Cycle Sequencing kit (Applied Biosystems, Waltham, MA, USA), according to the maker's instructions. The samples were sequenced in an ABI 3730 DNA analyzer (Applied Biosystems) installed at the DNA Sequencing Platform of the Oswaldo Cruz Institute (PDTIS/FIOCRUZ) in the RPT01A–DNA Sequencing subunit.

The chromatograms of the amplified sequences were assembled into contigs, analyzed and edited in Geneious Prime 2023.02.1 (http://www.geneious.com, accessed on 30 December 2023). This sequence was used to search GenBank (www.ncbi.nlm.nih.gov/genbank, accessed on 30 December 2023) for similar MT-CO1 sequences, using the BLAST (Basic Local Alignment Search Tool) in the BLASTn algorithm [41]. Sequences of *An. cantonensis* obtained from GenBank were used for the phylogenetic analyses, with four taxa of the genus *Angiostrongylus* being used as the outgroup, and one taxon of the genus *Aelurostrongylus*.

The MT-CO1 sequences were aligned using the Muscle tool, which was implemented in Geneious R9 [42] and the resulting matrix was edited to eliminate poorly aligned extremities and converted to the Nex format in Mesquite for the construction of the phylogenetic tree, version 3.51 [43]. The Markov chains were configured in the command block to be sampled at every 100 generations (sampleFreq = 100) in a total run of 10 million generations (ngen = 10,000,000). The posterior probabilities were calculated from the residuals, and a consensus sequence was generated based on the 50% majority rule.

Analyses of Bayesian Inference (BI) were run in MrBayes version 3.2.7 [44], using the GTR+I+G evolutionary model. The Bayesian analysis was run in the CIPRES Science Gateway V. 3.3 (https://www.phylo.org/, accessed on 30 December 2023) [45].

3. Results

In all five species of terrestrial mollusks were collected: *A. fulica* (*n* = 159), *Bulimulus tenuissimus* (d'Orbigny, 1835) (*n* = 35), *Subulina octona* (*n* = 19), and the slugs *Diplosolenodes occidentalis* (Guilding, 1825) (*n* = 29) and *Sarasinula linguaeformis* (Semper, 1885) (*n* = 64). The parasitological examination of 306 of the specimens collected revealed that 163 were parasitized by nematodes. Of these, 59 specimens presented angiostrongylid larvae from 29 of the 45 neighborhoods of Macapá (Figure 1 and Table 1). Except for *B. tenuissimus*, at least one specimen of each mollusk species was infected by angiostrongylid.

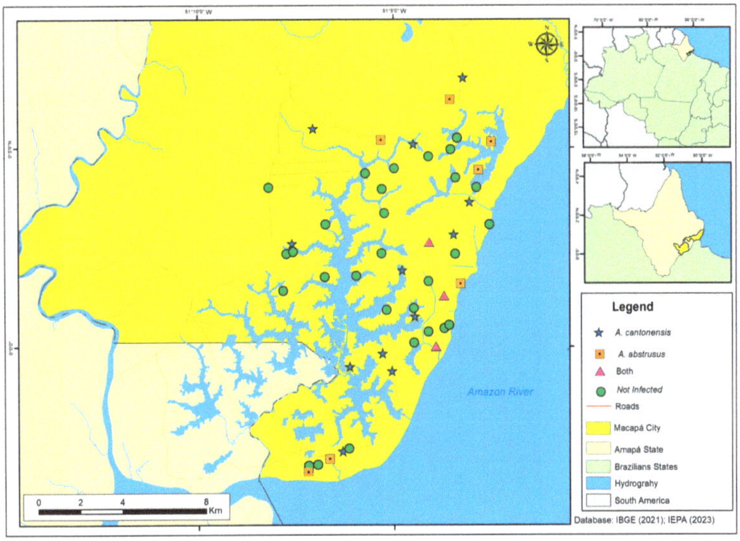

Figure 1. Study area in the municipality of Macapá, in northern Brazil, indicating the points from which mollusks were collected. Blue star: sites where the mollusks were positive for *An. cantonensis*; orange square: sites where mollusks were positive for *Ae. abstrusus*; green circle: site where the mollusks were not positive for Metastrongyloidea larvae; pink triangle: sites where both species were found, *An. cantonensis* and *Ae. abstrusus*.

Table 1. The neighborhoods of the municipality of Macapá in which mollusk specimens infected by angiostrongylid nematode larvae were collected during the present study. The values indicate the number of individuals infected by each type of nematode in each neighborhood. AA = *Aelurostrongylus abstrusus*, AC = *Angiostrongylus cantonensis*.

	Achatina fulica		*Subulina octona*	*Diplosolenodes occidentalis*	*Sarasinula linguaeformis*
Neighborhoods	**AA**	**AC**	**AC**	**AA**	**AC**
Amazonas					2
Beirol	1	4	2		
Boné Azul		2			
Embrapa		1			
Fazendinha	4				
Ipê	1				
Jardim Felicidade I	1			1	1
Jardim Felicidade II	4				
Jesus de Nazaré	1	1			
Julião Ramos		1			
Marabaixo III		3	1		2
Muca		1			
Murici	4				
Nova Esperança		2			
Pacoval		1			
Palmeiras	2				
Pedrinhas		3			
Renascer I	3				
Santa Inês	1				
São José	5				
Universidade		2			
Vale Verde		1			
Zerão		1			

The localities of Beirol and Marabaixo III showed greater epidemiological importance (seven and six snails parasitized by Metastrongyloidea, respectively), unlike the other three neighborhoods (Buritizal, Jardim Equatorial, Jardim Marco Zero, Novo Buritizal, Parque dos Buritis, and São Lázaro). Mollusks infected with An. cantonensis and Ae. abstrusus were collected at 15 and 11 localities, respectively (Table 1). Both nematode species were recorded in three neighborhoods (Beirol, Jardim Felicidade, and Jesus de Nazaré).

Infections of *An. cantonensis* were observed in *A. fulica* (23 specimens: 14.11% of the total of this species analyzed parasitologically), *S. linguaeformis* (five specimens: 7.81%), and *S. octona* (three specimens:15.78%); *Ae. abstrusus* was detected in only two species: *A. fulica* (27 specimens: 16.98%) and *D. occidentalis* (one: 3.44%) (Figures 2 and 3). The invasive exotic species *A. fulica* was the only mollusk parasitized by both nematodes (Table 1).

Figure 2. Photos of live specimens of the species found parasitized with Metastrongyloidea larvae in Macapá, Macapá, Brazilian Amazon Region: (**A**)—*Achatina fulica*, (**B**)—*Subulina octona*, (**C**)—*Diplosolenodes occidentalis*, (**D**)—*Sarasinula linguaeformis*.

Of the 163 infected mollusks, 36.19% were parasitized with *An. cantonensis* and *Ae. abstrusus* larvae, and the hosts were predominantly *A. fulica* (51.96% of the total snails collected in the area). Considering the 306 specimens obtained in Macapá, by species collected, the indices of infection by Metastrongyloidea (*An. cantonensis, Ae. abstrusus* and larvae not identified due to the low parasitic load) and other nematodes can be observed in Table 2. The terrestrial snail *B. tenuissimus* was represented by 35 specimens (11.43% of mollusks obtained in the area), with snails infected by only non-harmful nematodes (ten snails: 28.57% of the specimens of this species obtained in the area).

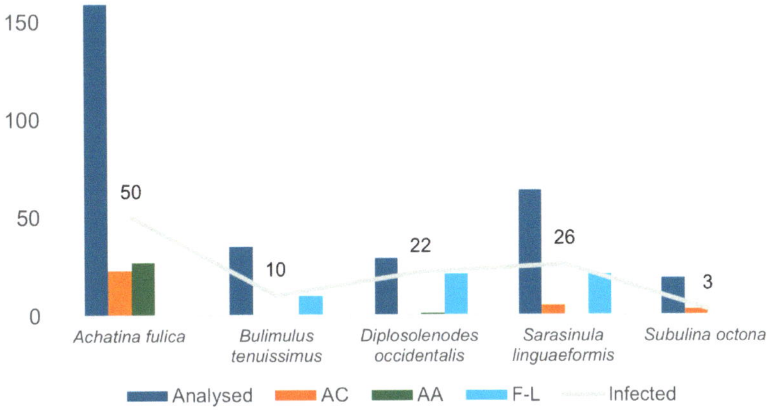

Figure 3. Results of the parasitological analysis of gastropods collected in the municipality of Macapá, in northern Brazil. AC: gastropods infected with *An. cantonensis*, AA: gastropods infected with *Ae. abstrusus*, F-L: free-living nematodes. Grey line indicated the total of infected specimens.

Table 2. Mollusk species infected by nematode larvae collected in Macapá. The values indicate the number of individuals infected (*n*) by each species of nematode and the percentage of infection considering the total number of parasitized mollusks.

Species	Collected (*n*)	Parasitized (*n*)	Metastrongyloidea, Angiostrongylidae			Other Nematodes (*n*)
			Angiostrongylus cantonensis Only (*n*)	*Aelurostrongylus abstrusus* Only (*n*)	Not Identificated	
A. fulica	159	108	23 (14.11%)	27 (16.56%)	10 (6.13%)	48 (29.45%)
B. tenuissimus	35	10	-	-		10 (6.13%)
D. occidentalis	29	21	-	1 (0.61%)		20 (12.27%)
S. linguaeformis	64	21	5 (3.06%)	-	4 (2.45%)	12 (7.36%)
S. octona	19	3	3 (1.84%)	-		-
Total	306	163				

The sequencing of the CO1 gene generated two good sequences (forward and reverse), resulting in 700 base pairs for each *An. cantonensis* sample. The present study generated 18 new sequences of *An. cantonensis* obtained from 14 specimens of *A. fulica*, three *S. linguaeformis*, and one *S. octona*. All the new sequences have been deposited in GenBank. In the phylogenetic analysis (Figure 4) we included 17 of these sequences (see Table 3) and three sequences from Amapá, previously published by Barbosa et al. [14] (GenBank accession numbers MN994436, MN994437, and MN994438). Some of the sequences were abnormally short and were not considered to be adequate for analysis. All the sequences recovered in the present study were of the same haplotype and were 99.5–100% similar to the *An. cantonensis* sequences deposited in GenBank.

The sequences obtained in the present study formed a distinct clade (BPP = 75%), which clustered with the *An. cantonensis* sequences from Australia, Brazil, Spain, and Taiwan. All the *An. cantonensis* sequences form a monophyletic group, albeit with low support, i.e., BPP = 53% (Figure 4). Monophyletic groups were observed among the different species of the same genus, such as *An. cantonensis* (BPP = 50%) and *Angiostrongylus mackerrasae* (Bhaibulaya, 1968) (BPP = 100%). The *An. cantonensis* sequence obtained in the present study from *S. octona* was excluded due to its abnormally short length.

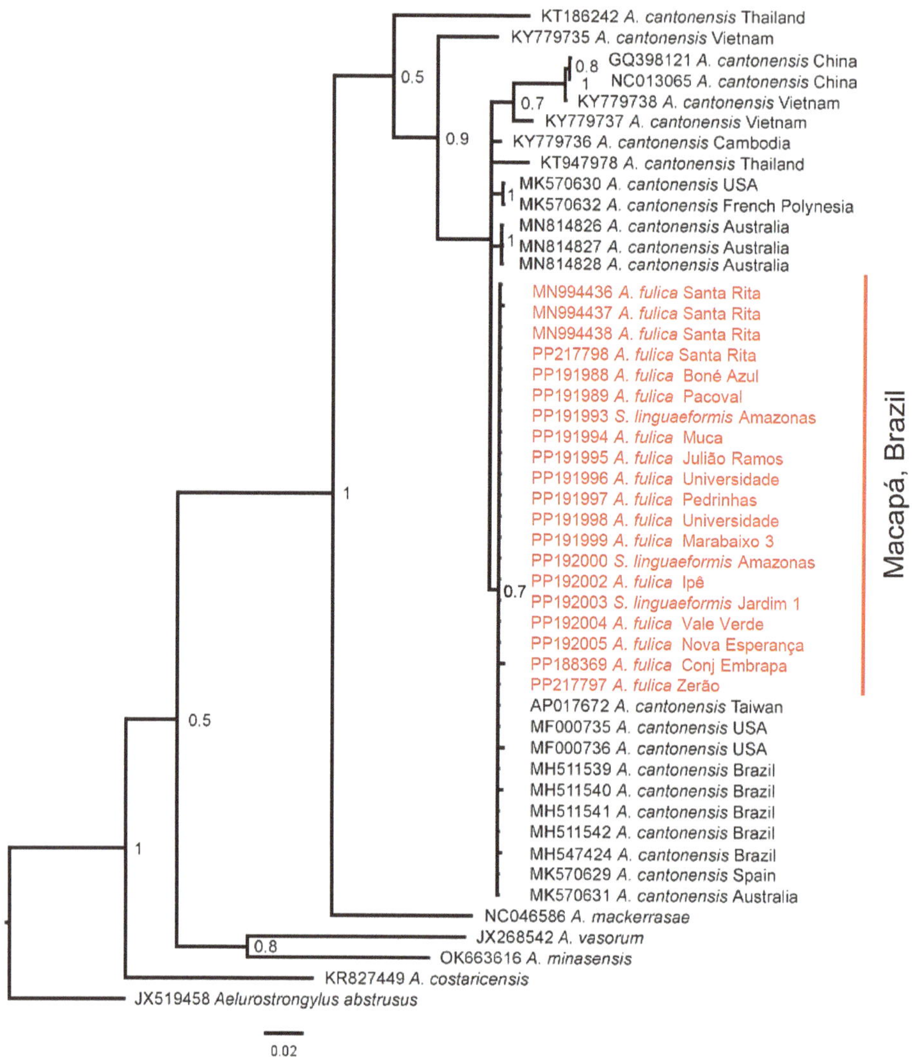

Figure 4. Phylogenetic reconstruction based on the partial sequences of the mitochondrial cytochrome *c* oxidase subunit 1 (CO1) gene using the Bayesian Inference (BI) approach. The 50% majority consensus tree obtained using the GTR + I + G model is shown. The numbers at the branch nodes are the Bayesian posterior probabilities of the 48 sequences, which include the 20 obtained from the present study, which are shown in pink, with their respective hosts in blue. *Aelurostrongylus abstrusus* was included as the outgroup.

Table 3. The nematode CO1 sequences included in the phylogenetic analysis of the present study, with their respective taxonomic classification, country, host, accession number, and reference.

Species	Geographic Location	Hosts	GenBank Acession Number	References
Angiostrongylus cantonensis	China	*Rattus norvegicus*	GQ398121	Lv et al. [46]
	China	*Rattus norvegicus*	NC013065	Lv et al. [46]
	Tailand	*Mus musculus*	KT947978	Yong et al. [47]
	USA	*Rattus exulans*	MK570630	Cervená et al. [48]
	French Polynesia	*Rattus exulans*	MK570632	Cervená et al. [48]
	Australia	*Rattus fuscipes*	MN814826	Valentyne et al. [49]
	Australia	*Rattus rattus*	MN814827	Valentyne et al. [49]
	Australia	*Rattus rattus*	MN814828	Valentyne et al. [49]
	Brazil	*Achatina fulica*	MN994436	Barbosa et al. [14]
	Brazil	*Achatina fulica*	MN994438	Barbosa et al. [14]
	Brazil	*Achatina fulica*	MN994437	Barbosa et al. [14]
	Brazil	*Achatina fulica*	PP188369	Present study
	Brazil	*Achatina fulica*	PP191988	Present study
	Brazil	*Achatina fulica*	PP191989	Present study
	Brazil	*S. linguaeformis*	PP191993	Present study
	Brazil	*Achatina fulica*	PP191994	Present study
	Brazil	*Achatina fulica*	PP191995	Present study
	Brazil	*Achatina fulica*	PP191996	Present study
	Brazil	*Achatina fulica*	PP191997	Present study
	Brazil	*Achatina fulica*	PP191998	Present study
	Brazil	*Achatina fulica*	PP191999	Present study
	Brazil	*S. linguaeformis*	PP192000	Present study
	Brazil	*Achatina fulica*	PP192002	Present study
	Brazil	*S. linguaeformis*	PP192003	Present study
	Brazil	*Achatina fulica*	PP192004	Present study
	Brazil	*Achatina fulica*	PP192005	Present study
	Brazil	*Achatina fulica*	PP217797	Present study
	Brazil	*Achatina fulica*	PP217798	Present study
	Taiwan	-	AP017672	Unpublish
	USA	*Didelphis virginiana*	MF000735	Dalton et al. [50]
	USA	*Didelphis virginiana*	MF000736	Dalton et al. [50]
	Brazil	*Achatina fulica*	MH511539	Ramos-de-Souza et al. [51]
	Brazil	*Achatina fulica*	MH511541	Ramos-de-Souza et al. [51]
	Brazil	*Cyclodontina fasciata*	MH511542	Ramos-de-Souza et al. [51]
	Brazil	*Bulimulus tenuissimus*	MH547424	Ramos-de-Souza et al. [51]
	Spain	*Rattus rattus*	MK570629	Cervená et al. [48]
	Australia	*Rattus rattus*	MK570631	Cervená et al. [48]
	Tailand	-	KT186242	Yong et al. [52]
	Cambodja	*Pomacea* sp.	KY779735	Lv et al. [53]
	Cambodja	*Pomacea* sp.	KY779736	Lv et al. [53]
	Vietnam	*Pomacea* sp.	KY779737	Lv et al. [53]
	Vietnam	*Pomacea* sp.	KY779738	Lv et al. [53]
Angiostrongylus mackerrasae	Australia	*Rattus fuscipes*	MN814821	Valentyne et al. [49]
	Australia	*Rattus fuscipes*	MN814822	Valentyne et al. [49]
	Australia	*Rattus fuscipes*	MN793157	Valentyne et al. [49]
	Australia	*Rattus fuscipes*	NC046586	Valentyne et al. [49]
Angiostrongylus malaysiensis	Malaysia	*Rattus rattus diardii*	KT947979	Yong et al. [47]
	Malaysia	*Rattus rattus diardii*	NC_030332	Yong et al. [47]
Angiostrongylus minasensis	Brazil	*Nasua nasua*	OK663616	Almeida et al. [4]
	Brazil	*Nasua nasua*	OK663635	Almeida et al. [4]
Angiostrongylus costaricensis	Costa Rica	*Nasua narica*	KX378965	Santoro et al. [54]
	Costa Rica	-	AP017675	Unpublish
	Costa Rica	-	KR827449	Yong et al. [55]

4. Discussion

Our results confirm and expand on the participation of the terrestrial mollusks in the maintenance of the life cycle of important metatrongyloids in Brazil, including *An. cantonensis* and *Ae. abstrusus.* Although Barbosa et al. [14] reported these nematodes in Macapá, the malacological investigation indicated only *A. fulica* acting in the transmission in the Santa Rita neighborhood, where the case of EM was reported in the municipality in 2018. Barbosa et al. [14] recorded *A. fulica* infected with *Ae. abstrusus* in the Santa Rita neighborhood of Macapá.

Both *An. cantonensis* and *Ae. abstrusus* have also been recorded in other regions of Brazil [3,21], as well as in other countries [2]. Carvalho et al. [22] reported the infection of a number of different mollusk species by *An. cantonensis* in areas adjacent to ports in different Brazilian states. Bechara et al. [56] also reported the occurrence of *A. fulica* and infected definitive hosts in anthropogenic areas of the municipality of Rio de Janeiro, in southeastern Brazil. Infestations of *A. fulica* are typically found on vacant lots, where refuse and decomposing organic material tends to accumulate [21,57,58], as observed at many of the sites in Macapá.

The *An. cantonensis* larvae obtained from the mollusks collected in Macapá and analyzed using CO1 sequences were 99.5~100% similar to the catalogued *An. cantonensis* sequences recovered in the BLAST search. In the phylogenetic tree obtained here, the *An. cantonensis* samples were closest to the other sequences obtained from Brazil, such as those identified in the state of Sergipe [51]. But they also clustered with sequences from Australia, Spain, and Taiwan, which indicates a closer relationship with *An. cantonensis* from these regions. A similar relationship between Brazilian *An. cantonensis* and populations from Asian countries was found by Monte et al. [59], who observed that the Brazilian sequences were close to those from Japan, China, and Thailand. These findings may reflect the proximity of ports at which ships carrying definitive (rodents) or intermediate hosts (mollusks) infected with *An. cantonensis* have docked [26,59–61]. The present study is the first to analyze samples of *An. cantonensis* from the Brazilian state of Amapá. Given the rapid dispersal of the snail *A. fulica* in Brazil and the fact that *An. cantonensis* has been found infecting these snails in 14 Brazilian states up to now [3,12,14,21,51,61,62], phylogenetic studies of this parasite may contribute to the understanding of the dynamics of its introduction to and dispersal in the country. *An. cantonensis* was also reported from other Brazilian states, i.e., Rio de Janeiro, Espírito Santo, Goiás, Mato Grosso, São Paulo, Sergipe, Minas Gerais [27,39], and Amazonas [63].

Rodrigues et al. [25] recently also found a strong association between this nematode and *A. fulica*, based on the parasitological analysis of mollusks collected from urban areas in 46 of the 92 municipalities of the state of Rio de Janeiro. These authors also recorded *Ae. abstrusus* in the veronicellid slug *Latipes erinaceus*. Penagos-Tabares et al. [23,24] also reported the infection of *Achatina fulica* by *Ae. abstrusus* in Colombia.

The present study also identified new associations between these nematodes and terrestrial mollusks in Macapá, specifically, the natural infection of the veronicellid slugs *S. linguaeformis* and *D. occidentalis* by *An. cantonensis* and *Ae. abstrusus*, respectively. Robinson et al. [64] reported that *D. occidentalis* occurs in the Lesser Antilles, Central America, and northern South America. The present study reports the first record of the species in Brazil as well as of its infection by *An. cantonensis*. By contrast, *An. cantonensis* has already been recorded infecting *S. linguaeformis* in the Brazilian states of São Paulo, Bahia, Espírito Santo, Pernambuco, and Pará [12,22].

A previous malacological research study revealed 21 species of exotic and native terrestrial molluscs to the city of Macapá. In the present study, an epidemiological investigation revealed the association of nematodes with some of these species of terrestrial mollusks [65]. *Achatina fulica* was the most common and widespread gastropod species in the study area. It was also the mollusk most infected by the nematodes, as well as the only species infected by both *An. cantonensis* and *Ae. abstrusus*. These findings further reinforce the importance, not only for the dispersal of *An. cantonensis*, but also the maintenance of its life cycle, in a number of different countries around the world [2,3,17,22].

While *B. tenuissimus* was not found to be infected naturally by metastrongyloid larvae in Macapá, Ramos-de-Souza et al. [51] recorded the association of *An. cantonensis* and this species in the Brazilian state of Sergipe. This mollusk is amply distributed in Brazil, where it has been recorded in eight states: Pará, Maranhão, Pernambuco, Bahia, Mato Grosso, Espírito Santo, Rio de Janeiro, and São Paulo [30]. In an experimental study designed to evaluate the potential of *B. tenuissimus* as an intermediate host, Martins et al. [66] found that its susceptibility to infection by *An. cantonensis* was 17.25%.

The highest nematode infection rates were recorded in the Beirol, Marabaixo III, and Pedrinhas neighborhoods, where the social and environmental conditions favor the proliferation of both mollusks and rodents. In northern Brazil, infected *Achatina fulica* had also been collected in the municipality of Barcelos, in the state of Amazonas, as well as in the city of Belém, in Pará state, together with infected specimens of *R. rattus* and *R. norvegicus* [67].

Exotic species often have a disproportionate impact on the environment, given their ecological plasticity, lack of natural predators, and their potential for the transfer of parasites to the native fauna, a process known as "spillover" [2]. The results of the present study indicate that the giant African land snail *Achatina fulica* is a typical example of this problem, given its ample dispersal throughout all the states in Brazil, as well as the Federal District, where it is abundant in many urban areas, especially where refuse and debris accumulate, which favors the presence of rodents and the completion of the life cycle of *An. cantonensis* and other zoonotic nematodes [3,12].

Webster et al. [68] highlighted the One Health concept, which associates the health of human populations with that of both animals and the environment, with a focus on the transmission of zoonotic diseases. Zinsstag et al. [69] also reinforced the importance of the One Health concept, which they associate with the term One Medicine, which emphasizes the need for the integration and convergence of the health of all the species and the ecosystems they inhabit. Woldehanna and Zimicki [70] also pointed out the importance of identifying the relationship between human behavior and the probable infection routes, to ensure the development of effective preventive measures. In particular, the social dynamic determines the level of possible interactions with animals, as well as the intensity of these interactions and, in turn, the potential for exposure to pathogens.

5. Conclusions

The present study demonstrated the ample distribution of *An. cantonensis* in the municipality of Macapá and the maintenance of its life cycle through the presence of different species of terrestrial mollusk, besides *A. fulica*. It also provides the first record of the occurrence of the slug *Diplosolenodes occidentalis* infected by *Ae. abstrusus* in Brazil, as well as the first records of the natural infection of *Subulina octona* and *Sarasinula linguaeformis* by *An. cantonensis* in the Brazilian Amazon region.

The finding of different species of mollusks infected with parasitic nematodes highlights the importance of such studies for public and veterinary health. We also emphasize the need for the implementation of measures to improve the health of the local population, in particular through health education, and the surveillance and control of the mollusk intermediate hosts. These measures are fundamental to the prevention of the transmission of neglected zoonotic diseases, in particular EM, in urban areas where infected mollusks are relatively abundant.

Author Contributions: T.A.B., S.C.T., M.A.F. and S.R.G. planned and implemented the field expeditions in the urban zone of Macapá, the morphological and parasitological analyses; J.R.-d.-S. performed the molecular and phylogenetic analyzes. All authors performed the literature research, wrote and reviewed the manuscript. All authors have read and agreed to the published version of the manuscript.

Funding: The study was supported by internal funding of the National Reference Laboratory for Schistosomiasis–Malacology (LRNEM), Oswaldo Cruz Institute (IOC), considering a case of eosinophilic meningitis in humans reported in this municipality in 2018, and own funds of the first author.

Institutional Review Board Statement: Not applicable.

Data Availability Statement: Mollusks samples were deposited in the Mollusk Collection of the Oswaldo Cruz Institute (CMIOC/Fiocruz) in Rio de Janeiro. The database of this collection can be acessed in the following address specieslink.net. The sequences of *An. cantonensis* generated in this study were deposited in GenBank GenBank Overview (nih.gov).

Acknowledgments: We are grateful to the LMALAC/LRNEM staff for scientific and technological support. We would also like to thank Marta Chagas Pinto for her assistance with parasitological examination of snails by the digestion technique and Arielly Kelly Porfírio de Sousa for assistance with the DNA processing. We are also grateful to the IEPA for provide the facilities for the sorting and handling of the material collected the field and Orleno Marques for producing the map. We thank Elivaldo M. Nunes for his help collecting the mollusks, and Fred Julio C. Monteiro, head of UCV-LACEN (2019) for sending the samples to LRNEM.

Conflicts of Interest: The authors declare no conflicts of interest.

References

1. Prociv, P.; Spratt, D.M.; Carlisle, M.S. Neuro-angiostrongyliasis: Unresolved issues. *Int. J. Parasitol.* **2000**, *30*, 1295–1303. [CrossRef] [PubMed]
2. Valente, R.; Robles, M.R.; Diaz, J.I. Gastropods as intermediate hosts of *Angiostrongylus* spp. in the Americas: Bioecological characteristics and geographical distribution. *Mem. Inst. Oswaldo Cruz* **2020**, *115*, e200236. [CrossRef] [PubMed]
3. Thiengo, S.C.; Ramos-de-Souza, J.; Silva, G.M.; Fernandez, M.A.; Silva, E.F.; Sousa, A.K.P.; Rodrigues, P.S.; Mattos, A.C.; Costa, R.A.F.; Gomes, S.R. Parasitism of terrestrial gastropods by medically-important nematodes in Brazil. *Front. Vet. Sci.* **2022**, *9*, 1023426. [CrossRef] [PubMed]
4. Almeida, L.R.; Souza, J.G.R.; Santos, H.A.; Torres, E.J.L.; Vilela, R.V.; Cruz, O.M.S.; Rodrigues, L.; Pereira, C.A.J.; Maldonado Junior, A.; Lima, W.S. *Angiostrongylus minasensis* n. sp.: New species found parasitizing coatis (*Nasua nasua*) in an urban protected area in Brazil. *Rev. Bras. Parasitol. Vet.* **2020**, *29*, e018119. [CrossRef]
5. Neves, D.P. *Parasitologia Humana*, 13th ed.; Atheneu: São Paulo, Brazil, 2016.
6. Rey, L. *Parasitologia: Parasitos e Doenças Parasitárias do Homem nos Trópicos Ocidentais*, 4th ed.; Guanabara Koogan: Rio de Janeiro, Brazil, 2018.
7. Spratt, D.M. Species of *Angiostrongylus* (Nematoda: Metastrongyloidea) in wildlife: A review. *Int. J. Parasitol. Parasites Wildl.* **2015**, *4*, 178–189. [CrossRef]
8. Caldeira, R.L.; Mendonça, C.L.G.F.; Gouveia, C.O.; Lenzi, H.L.; Graeff-Teixeira, C.; Lima, W.S.; Mota, E.M.; Pecora, I.L.; de Medeiros, A.M.Z.; Carvalho, O.D.S. First record of molluscs naturally infected with *Angiostrongylus cantonensis* (Chen, 1935) (Nematoda: Metastrongylidae) in Brazil. *Mem. Inst. Oswaldo Cruz* **2007**, *102*, 887–889. [CrossRef]
9. Kim, J.R.; Hayes, K.A.; Yeung, N.W.; Cowie, R.H. Diverse Gastropod Hosts of *Angiostrongylus cantonensis*, the Rat Lungworm, Globally and with a Focus on the Hawaiian Islands. *PLoS ONE* **2014**, *9*, e94969. [CrossRef]
10. Lima, A.R.M.C.; Mesquita, S.D.; Santos, S.S.; Aquino, E.R.P.D.; Rosa, L.R.S.; Duarte, F.S.; Teixeira, A.O.; da Silva Costa, Z.R.; Ferreira, M.L.B. Alicata disease: Neuroinfestation by *Angiostrongylus cantonensis* in Recife, Pernambuco, Brazil. *Arq. Neuropsiquiatr.* **2009**, *67*, 1093–1096. [CrossRef]
11. Thiengo, S.C.; Maldonado, A.; Mota, E.M.; Torres, E.J.L.; Caldeira, R.; Carvalho, O.D.S.; Oliveira, A.P.M.; Simões, R.O.; Fernandez, M.A.; Lanfredi, R.M. The giant African snail *A. fulica* as natural intermediate host of *A. cantonensis* in Pernambuco, northeast Brazil. *Acta Trop.* **2010**, *115*, 194–199. [CrossRef]
12. Morassoutti, A.L.; Thiengo, S.C.; Fernandez, M.; Sawanyawisuth, K.; Graeff-Teixeira, C. Eosinophilic meningitis caused by *Angiostrongylus cantonensis*: Na emergente disease in Brazil. *Mem. Inst. Oswaldo Cruz* **2014**, *109*, 399–407. [CrossRef]
13. Valente, R.; Robles, M.R.; Navone, G.T.; Diaz, J.I. *Angiostrongylus* spp. in the Americas: Geographical and chronological distribution of definitive hosts versus disease reports. *Mem. Inst. Oswaldo Cruz* **2018**, *113*, 143–152. [CrossRef] [PubMed]
14. Barbosa, T.A.; Thiengo, S.C.; Fernandez, M.A.; Graeff-Teixeira, C.; Morassuti, A.L.; Mourão, F.R.P.; Miranda, C.O.S.; de Moraes Jorge, M.; Costa, L.F.; Gomes, S.R. Infection by *Angiostrongylus cantonensis* in both humans and the snail *Achatina* (*Lissachatina*) *fulica* in the city of Macapá, in the Amazon Region of Brazil. *Mem. Inst. Oswaldo Cruz* **2020**, *115*, e200115. [CrossRef] [PubMed]
15. Paiva, J.A.C.D.; Ferreira, J.D.B.; Santos, R.W.V. Primeiro caso de meningite eosinofílica causada por *Angiostrongylus cantonensis* no estado do Amapá. *Rev. Cient. Multidiscip. Núcl. Conhecimento* **2020**, *10*, 124–132. [CrossRef]
16. Espírito-Santo, M.C.C.; Pinto, P.L.S.; Mota, D.J.G.D. The first case of *Angiostrongylus cantonensis* eosinophilic meningitis diagnosed in the city of São Paulo, Brazil. *Rev. Inst. Med. Trop. São Paulo* **2013**, *55*, 129–132. [CrossRef] [PubMed]

17. Giraldo, A.; Garzòn, C.; Castillo, A.; Córdoba-Rojas, D.F. Confirmation of the presence of *Angiostrongylus cantonensis* in lung tissue of the African giant snail (*Lissachatina fulica*) in Colombia. *Infectio* **2019**, *23*, 129–132. [CrossRef]
18. Rollins, R.L.; Medeiros, M.C.I.; Cowie, R.H. Stressed snails release *Angiostrongylus cantonensis* (rat lungworm) larvae in their slime. *One Health* **2023**, *17*, 100658. [CrossRef] [PubMed]
19. Maldonado, A., Jr.; Simões, R.; Thiengo, S. Angiostrongyliasis in the Americas. *J. Zoonosis Intech* **2012**, *10*, 38632.
20. Simões, R.O.; Maldonado, A., Jr.; Olifiers, N.; Garcia, J.S.; Bertolino, A.V.; Luque, J.L. A longitudinal study of *Angiostrongylus cantonensis* in an urban population of *Rattus norvegicus* in Brazil: The influences of seasonality and host features on the pattern of infection. *Parasites Vectors* **2014**, *7*, 100. [CrossRef]
21. Silva, G.M.; Thiengo, S.C.; Menezes, A.N.; Melo, C.M.; Jeraldo, V.L. Relative condition factor and predictive model for the presence of the invasive snail *Achatina (Lissachatina) fulica* in Sergipe, Northeast Brazil. *Biota Neotrop.* **2022**, *22*, e20211323. [CrossRef]
22. Carvalho, O.S.; Sholte, R.G.C.; Mendonça, C.L.F.; Passos, L.K.J.; Caldeira, R.L. *Angiostrongylus cantonensis* (Nematode: Metastrongyloidea) in molluscs from harbour areas in Brazil. *Mem. Inst. Oswaldo Cruz* **2012**, *107*, 740–746. [CrossRef]
23. Penagos-Tabares, F.; Lange, M.K.; Chaparro-Gutiérrez, J.J.; Taubert, A.; Hermosilla, C. *Aelurostrongylus abstrusus* and *Angiostrongylus vasorum*: Neglected and underestimated parasites in South America. *Parasites Vectors* **2018**, *11*, 208. [CrossRef] [PubMed]
24. Penagos-Tabares, F.; Lange, M.K.; Vélez, J.; Hirzmann, J.; Gutiérrez-Arboleda, J.; Taubert, A.; Hermosilla, C.; Gutiérrez, J.J.C. The invasive giant African snail *Lissachatina fulica* as natural intermediate host of *Aelurostrongylus abstrusus*, *Angiostrongylus vasorum*, *Troglostrongylus brevior* and *Crenosoma vulpis* in Colombia. *PLoS Negl. Trop. Dis.* **2019**, *13*, e0007277. [CrossRef] [PubMed]
25. Rodrigues, P.S.; Gomes, S.R.; Montresor, L.C.; Ramos-de-Souza, J.; Barros, L.A.; Fernandez, M.A.; Thiengo, S.C. The giant African snail *Achatina (Lissachatina) fulica* Bowdich, 1822 as an intermediate host of *Aelurostrongylus abstrusus* (Railliet, 1898) in the Rio de Janeiro state, Brazil. *Vet. Parasitol. Reg. Stud. Rep.* **2022**, *30*, 100712. [CrossRef] [PubMed]
26. Valente, R.; Diaz, J.I.; Salomón, O.D.; Navone, G.T. Natural infection of the feline lungworm *Aelurostrongylus abstrusus* in the invasive snail *Achatina fulica* from Argentina. *Vet. Parasitol.* **2017**, *235*, 17–19. [CrossRef] [PubMed]
27. Thiengo, S.C.; Faraco, F.A.; Salgado, N.C.; Cowie, R.H.; Fernandez, M.A. Rapid spread of an invasive snail in South America: The giant African snail, *Achatina fulica*, in Brazil. *Biol. Invasions* **2007**, *9*, 693–702. [CrossRef]
28. Araújo, J.D.B.; Bessa, E.D.A. Moluscos de importância econômica no Brasil. II Subulinidae, *Subulina octona* (Bruguiére) (Mollusca, Gastropoda, Pulmonata, Stylommatophora). *Rev. Bras. Zool.* **1993**, *10*, 489–497. [CrossRef]
29. Bouchet, P.; Rocroi, J.P.; Hausdorf, B.; Kaim, A.; Kano, Y.; Nützel, A.; Parkhaev, P.; Schrödl, M.; Strong, E.E. Revised classification, nomenclator and typification of gastropod and monoplacophoran families. *Malacologia* **2017**, *61*, 1–526. [CrossRef]
30. Simone, L.R.L. *Land and Freshwater Molluscs of Brazil*, 1st ed.; EGB: São Paulo, Brazil, 2006; p. 390.
31. Thomé, J.W. Erneute Beschreibung neotropischer Veronicellidae-Typen (Mollusca, Gastropoda). *Steenstrupia* **1973**, *3*, 31–50.
32. Thomé, J.W. Os gêneros da família Veronicellidae nas Américas (Mollusca; Gastopoda). *Iheringia* **1975**, *48*, 3–56.
33. Thomé, J.W. Redescrição dos tipos de Veronicellidae (Mollusca, Gastropoda) Neotropicais. X. Os tipos de *Diplosolenodes occidentalis* (Guilding, 1825) no British Museum (Natural History), Londres. *Rev. Bras. Zool.* **1985**, *2*, 411–417. [CrossRef]
34. Thomé, J.W. Annotated and illustrated preliminary list of the Veronicellidae (Mollusca: Gastropoda) of the Antilles, and Central and North America. *J. Med. Appl. Malacol.* **1989**, *1*, 11–28.
35. Thomé, J.W.; Gomes, S.R.; Picanço, J.B. *Os Caracóis e as Lesmas dos Nossos Bosques e Jardins: Guia Ilustrado*; USEB: Pelotas, Brazil, 2006; ISBN 979-8-58998-510-8.
36. Graeff-Teixeira, C.; Morera, P. Método de digestão de moluscos em ácido clorídrico para isolamento de larvas de Metastrongilídeos. *Biociências* **1995**, *3*, 85–89.
37. Wallace, C.D.; Rosen, L. Techniques for recovering and identifying larvae of *Angiostrongylus cantonensis* from molluscs. *Malacologia* **1969**, *7*, 427–438.
38. Ash, L.R. Diagnostic morphology of the third-stage larvae of *Angiostrongylus cantonensis*, *Angiostrongylus vasorum*, *Aelurostrongylus abstrusus*, and *Anafilaroides rostratus* (Nematoda: Metastrongyloidea). *J. Parasitol.* **1970**, *56*, 249–253. [CrossRef] [PubMed]
39. Thiengo, S.C.; Fernandez, M.A.; Torres, E.J.; Coelho, P.M.; Lanfredi, R.M. First record of a nematode Metastrongyloidea (*Aelurostrongylus abstrusus* larvae) in *Achatina (Lissacathina) fulica* (Mollusca: Achatinidae) in Brazil. *J. Invertebr. Pathol.* **2008**, *98*, 34–39. [CrossRef] [PubMed]
40. Prosser, S.W.; Velarde-Aguilar, M.G.; León-Règagnon, V.; Hebert, P.D.N. Advancing nematode barcoding: A primer cocktail for the cytochrome c oxidase subunit I gene from vertebrate parasitic nematodes. *Mol. Ecol. Resour.* **2013**, *13*, 1108–1115. [CrossRef] [PubMed]
41. Altschul, S.F.; Gish, W.; Miller, W.; Myers, E.W.; Lipman, D.J. Basic local alignment search tool. *J. Mol. Biol.* **1990**, *215*, 403–410. [CrossRef] [PubMed]
42. Kearse, M.; Moir, R.; Wilson, A.; Stones-Havas, S.; Cheung, M.; Sturrock, S.; Buxton, S.; Cooper, A.; Markowitz, S.; Duran, C.; et al. Geneious Basic: An integrated and extendable desktop software platform for the organization and analysis of sequence data. *Bioinformatics* **2012**, *28*, 1647–1649. [CrossRef]
43. Maddison, W.P.; Maddison, D.R. Mesquite: A Modular System for Evolutionary Analysis. Version 3.61. 2018. Available online: http://www.mesquiteproject.org (accessed on 20 February 2019).
44. Ronquist, F.; Teslenko, M.; van der Mark, P.; Ayres, D.L.; Darling, A.; Höhna, S.; Larget, B.; Liu, L.; Suchard, M.A.; Huelsenbeck, J.P. MrBayes 3.2: Efficient Bayesian phylogenetic inference and model choice across a large model space. *Syst. Biol.* **2012**, *61*, 539–542. [CrossRef]

45. Miller, M.A.; Pfeiffer, W.; Schwartz, T. Creating the CIPRES Science Gateway for Inference of Large Phylogenetic Trees. In Proceedings of the Gateway Computing Environments Workshop (GCE), New Orleans, LA, USA, 14 November 2010. Available online: https://ieeexplore.ieee.org/document/5676129 (accessed on 20 February 2019).
46. Lv, S.; Zhang, Y.; Liu, H.-X.; Hu, L.; Liu, Q.; Wei, F.-R.; Guo, Y.-H.; Steinmann, P.; Hu, W.; Zhou, X.-N.; et al. Phylogenetic evidence for multiple and secondary introductions of invasive snails: *Pomacea* species in the People's Republic of China. *Divers. Distrib.* **2013**, *19*, 147–156. [CrossRef]
47. Yong, H.S.; Song, S.L.; Eamsobhana, P.; Lim, P.E. Complete mitochondrial genome of *Angiostrongylus malasiensis* lungworm and molecular phylogeny of Metastrongloid nematodes. *Acta Trop.* **2016**, *161*, 33–40. [CrossRef]
48. Cervená, B.; Modrý, D.; Fecková, B.; Hrazdilová, K.; Foronda, P.; Alonso, A.M.; Lee, R.; Walker, J.; Niebuhr, C.N.; Malik, R.; et al. Low diversity of *Angiostrongylus cantonensis* complete mitochondrial DNA sequences from Austrália, Hawaii, French Polynesia and the Canary Islands revealed using whole genome net-generation sequencing. *Parasites Vectors* **2019**, *12*, 241. [CrossRef] [PubMed]
49. Valentyne, H.; Spratt, D.M.; Aghazadeh, M.; Jones, M.K.; Šlapeta, J. The mitochondrial genome of *Angiostrongylus mackerrasae* is distinct from *A. cantonensis* and *A. malaysiensis. Parasitology* **2020**, *147*, 681–688. [CrossRef] [PubMed]
50. Dalton, M.F.; Fenton, H.; Cleveland, C.A.; Elsmo, E.J.; Yabsley, M.J. Eosinophilic meningoencephalitis associated with rat lungworm (*Angiostrongylus cantonensis*) migration in two-nine-banded armadillos (*Dasypus novemcinctus*) and an opossum (*Didelphis virginiana*) in the southeastern United States. *Int. J. Parasitol. Parasites Wildl.* **2017**, *6*, 131–134. [CrossRef] [PubMed]
51. Ramos-de-Souza, J.; Thiengo, S.C.; Fernandez, M.A.; Gomes, S.R.; Antônio, J.C.; Clímaco, M.C.; Garcia, J.S.; Maldonado-Junior, A.; Barbosa, L.; Dolabella, S.S. First records of molluscs naturally infected with *Angiostrongylus cantonensis* (Nematoda: Metastrongyloidea) in Sergipe state Northeastern Brazil, including new global records of natural intermediate hosts. *Rev. Inst. Med. Trop.* **2018**, *60*, e51. [CrossRef] [PubMed]
52. Yong, H.S.; Song, S.L.; Eamsobhana, P.; Goh, S.Y.; Lim, P.E. Complete mitochondrial genome reveals genetic diversity of *Angiostrongylus cantonensis* (Nematoda: Angiostrongylidae). *Acta Trop.* **2015**, *152*, 157–164. [CrossRef] [PubMed]
53. Lv, S.; Guo, Y.H.; Nguyen, H.M.S.; Sayasone, S.; Lo, N.C.; Zhou, X.N.; Andrews, J.R. Invasive Pomacea snails as important intermediate hosts of *Angiostrongylus cantonensis* in Laos, Cambodia and Vietnam: Implications for outbreaks of eosinophilic meningitis. *Acta Trop.* **2018**, *183*, 32–35. [CrossRef] [PubMed]
54. Santoro, M.; Alfaro-Alarcón, A.; Veneziano, V.; Cerrone, A.; Latrofa, M.S.; Otranto, D.; Hagnauer, I.; Jiménez, M.; Galiero, G. The White-nosed coati (*Nasua narica*) is a naturally susceptible definitive host for the zoonotic nematode *Angiostrongylus costaricensis* in Costa Rica. *Vet. Parasitol.* **2016**, *228*, 93–95. [CrossRef]
55. Yong, H.S.; Eamsobhana, P.; Song, S.L.; Prasartvit, A.; Lim, P.E. Molecular phylogeography of *Angiostronglus cantonensis* (Nematoda: Angiostrongylidae) and genetic relationships with congeners using cytochrome b gene marker. *Acta Trop.* **2015**, *148*, 66–71. [CrossRef]
56. Bechara, A.H.; Simões, R.O.; Faro, M.J.; Garcia, J.S. *Achatina fulica* infected by *Angiostrongylus cantonensis* on beaches, in the west zone of Rio de Janeiro, Brazil. *Rev. Inst. Med. Trop. São Paulo* **2018**, *60*, e4. [CrossRef]
57. Fischer, M.L.; Colley, E.; Amadigi, S.; Simião, M.S. O caracol gigante africano *Achatina fulica* no Brasil. In *Ecologia de Achatina fulica;* Champagnat: Curitiba, Brazil, 2010; pp. 101–140.
58. Silva, G.M.; Santos, M.B.; Melo, C.M.; Jeraldo, V.L. *Achatina fulica* (Gastropoda: Pulmonata): Ocorrência, aspectos ambientais e presença de nematódeos em Sergipe, Brasil. *Braz. J. Biol.* **2020**, *80*, 245–254. [CrossRef] [PubMed]
59. Monte, T.C.C.; Simões, R.O.; Oliveira, A.P.M.; Novaes, C.F.; Thiengo, S.C.; Silva, A.J.; Estrela, P.C.; Júnior, A.M. Phylogenetic relationship of the Brazilian isolates of the rat lungworm *Angiostrongylus cantonensis* (Nematoda: Metastrongylidae) employing mitochondrial COI gene sequence data. *Parasites Vectors* **2012**, *5*, 248. [CrossRef] [PubMed]
60. Maldonado Júnior, A.; Simões, R.O.; Oliveira, A.P.M.; Motta, E.M.; Fernandez, M.A.; Pereira, Z.M.; Monteiro, S.S.; Torres, E.J.L.; Thiengo, S.C. First report of *Angiostrongylus cantonensis* (Nematoda: Metastrongylidae) in *Achatina fulica* (Mollusca: Gastropoda) from Southeast and South Brazil. *Mem. Inst. Oswaldo Cruz* **2010**, *105*, 938–941. [CrossRef]
61. Ramos-de-Souza, J.; Gomes, S.R.; Mattos, A.C.; Sousa, A.K.P.; Silva, E.F.; Maldonado-Junior, A.; Thiengo, S.C. *Achatina fulica* infected by *Angiostrongylus cantonensis* in Manaus, brazilian amazon region, and the risk of transmission of eosinophilic meningitis. *J. Trop. Pathol.* **2023**, *52*, 295–303. [CrossRef]
62. Almeida, L.R.; Souza, J.J.; Botelho, L.M.; Vidigal, T.H.D.A.; Ecco, R.; Trindade, G.S.; Paglia, A.P.; de Jesus Pereira, C.A.; Lima, W.D.S. Parasitism in *Rattus rattus* and sympatric *Achatina fulica* by *Angiostrongylus cantonensis* in an urban park in southeast Brazil. *Parasitol. Resour.* **2022**, *122*, 347–352. [CrossRef] [PubMed]
63. Andrade-Porto, S.M.; Souza, K.C.P.; Cárdenas, M.Q.; Roque, R.A.; Pimpão, D.M.; Araújo, C.S.; de Oliveira Malta, J.C. Occurrence of *Aelurostrongylus abstrusus* (Railliet, 1898) larvae (Nematoda: Metastrongylidae) infecting *Achatina* (*Lissachatina*) *fulica* Bowdich, 1822 (Mollusca: Gastropoda) in the Amazon region. *Acta Amaz.* **2012**, *42*, 245–250. [CrossRef]
64. Barbosa, T.A.; Fernandez, M.A.; Thiengo, S.C.; Gomes, S.R. Criação da coleção de molusco: Uma contribuição ao estudo da malacofauna Amapaense. In *Coleções Científicas do Amapá;* Patrick, C., Cecile, G., Lúcio, S.C.L., Eds.; Núcleo de Informação e Documentação—NUID/IEPA: Macapá, Brazil, 2021; Volume I, pp. 107–117.
65. Robinson, D.G.; Hovestadt, A.; Fields, A.; Breure, A. The land Mollusca of Dominica (Lesser Antilles) with notes some enigmatic or rare species. *Zoo Med.* **2009**, *83*, 615–650.

66. Martins, F.G.; Garcia, J.S.; Torres, E.J.L.; Santos, M.A.J.; Massard, C.L.; Pinheiro, J. Firt record of *Bulimulus tenuissimus* (Mollusca) as potential experimental intermediate host of *Angiostrongylus cantonensis* (Nematode). *Braz. J. Biol.* **2019**, *79*, 686–696. [CrossRef]
67. Moreira, V.; Giese, E.; Melo, F.; Simões, R.; Thiengo, S.; Maldonado Junior, A.; Santos, J.N. Endemic angiostrongyliasis in the Brazilian Amazon: Natural parasitism of *Angiostrongylus cantonensis* in *Rattus rattus* and *R. norvegicus*, and sympatric giant African land snails, *Achatina fulica*. *Acta Trop.* **2013**, *125*, 90–97. [CrossRef]
68. Webster, J.P.; Gower, C.M.; Knowles, S.C.L.; Molyneux, D.H.; Fenton, A. One health–an ecological and evolutionary framework for tackling Neglected Zoonotic Diseases. *Evol. Appl.* **2016**, *9*, 313–333. [CrossRef]
69. Zinsstag, J.; Schelling, E.; Waltner-Toews, D.; Tanner, M. From "one medicine" to "one health" and systemic approaches to health and well-being. *Prev. Vet. Med.* **2011**, *101*, 148–156. [CrossRef]
70. Woldehanna, S.; Zimiki, S. An expanded One Health model: Integrating social science and One Health to inform study of the human-animal interface. *Soc. Sci. Med.* **2015**, *129*, 87–95. [CrossRef]

pathogens

MDPI

Article

Parasite Fauna and Coinfections in Urban Rats Naturally Infected by the Zoonotic Parasite *Angiostrongylus cantonensis*

María Teresa Galán-Puchades [1,*], Carla Gosálvez [1], María Trelis [1], Mercedes Gómez-Samblás [2], Jennifer Solano-Parada [2], Antonio Osuna [2], Sandra Sáez-Durán [1], Rubén Bueno-Marí [1,3] and Marius V. Fuentes [1]

[1] Parasites & Health Research Group, Department of Pharmacy, Pharmaceutical Technology and Parasitology, Faculty of Pharmacy, University of Valencia, Burjassot, 46100 Valencia, Spain; carla.go.can@gmail.com (C.G.); maria.trelis@uv.es (M.T.); sandra.saez@uv.es (S.S.-D.); ruben.bueno@uv.es (R.B.-M.); mario.v.fuentes@uv.es (M.V.F.)

[2] Laboratory of Biochemistry and Molecular Parasitology, Institute of Biotechnology, University of Granada, 18071 Granada, Spain; msambla@gmail.com (M.G.-S.); jesolpa@gmail.com (J.S.-P.); aosuna@ugr.es (A.O.)

[3] Laboratorios Lokímica, Departamento de Investigación y Desarrollo (I+D), Ronda Auguste y Louis Lumière 23, Nave 10, Parque Tecnológico, Paterna, 46980 Valencia, Spain

* Correspondence: mteresa.galan@uv.es

Abstract: When the zoonotic parasite of rodents that can cause human neuroangiostrongyliasis, i.e., *Angiostrongylus cantonensis,* is found in its natural definitive hosts, it is usually reported in isolation, as if the rat lungworm were the only component of its parasite community. In this study, we report the coinfections found in rats naturally infected by *A. cantonensis* in urban populations of *Rattus norvegicus* and *Rattus rattus* in Valencia, Spain. In addition to the rat lungworms, which were found in 14 of the 125 rats studied (a prevalence of 11.20%), 18 other parasite species (intestinal and tissular protists, microsporidia and helminths) were found, some of them with high burdens. Fourteen of these nineteen species found are potential zoonotic parasites, namely *Blastocystis, Giardia duodenalis, Cryptosporidium* spp., *Enterocytozoon bieneusi, Encephalitozoon hellem, Toxoplasma gondii, Brachylaima* spp., *Hydatigera taeniaeformis s.l.* larvae, *Hymenolepis nana, Hymenolepis diminuta, Angiostrongylus cantonensis, Calodium hepaticum, Gongylonema neoplasticum* and *Moniliformis moniliformis.* The total predominance of coinfected rats as well as their high parasite loads seem to indicate a trend towards parasite tolerance.

Keywords: *Angiostrongylus cantonensis*; *Rattus norvegicus*; *Rattus rattus*; parasite fauna; coinfections

Citation: Galán-Puchades, M.T.; Gosálvez, C.; Trelis, M.; Gómez-Samblás, M.; Solano-Parada, J.; Osuna, A.; Sáez-Durán, S.; Bueno-Marí, R.; Fuentes, M.V. Parasite Fauna and Coinfections in Urban Rats Naturally Infected by the Zoonotic Parasite *Angiostrongylus cantonensis*. *Pathogens* **2024**, *13*, 28. https://doi.org/10.3390/pathogens13010028

Academic Editor: Susan Jarvi

Received: 9 November 2023
Revised: 19 December 2023
Accepted: 25 December 2023
Published: 27 December 2023

1. Introduction

Angiostrongylus cantonensis (Chen, 1935) (Rhabditida: Angiostrongylidae) is a zoonotic parasite of rodents, including mainly the Norway rat, *Rattus norvegicus*, and the black rat, *Rattus rattus*, as their natural definitive hosts, which become infected through the ingestion of its intermediate hosts, snails and slugs, or its paratenic hosts, such as freshwater prawns, frogs and land crabs, among others [1]. The parasite is known as the rat lungworm, as the adults live in the pulmonary arteries of rats. When humans accidentally become infected—via the same route of transmission as rats—the parasite can cause neuroangiostrongyliasis due to the presence of the worms in the central nervous system [1]. Therefore, its control must be established under the One Health concept, as the control of zoonoses is an integral part of this approach.

Angiostrongylus cantonensis has been reported mainly in tropical and subtropical areas, limited by low temperatures. Until recently, the parasite seemed to be far away from Europe. However, it was found in rats and snails in Tenerife (Canary Islands, Spain) in 2010 [2]. Yet, although Tenerife is Europe in a political sense, it is Africa, geographically speaking. Several years later, in 2019, the parasite was found in hedgehogs in Mallorca (Balearic Islands, Spain) which, geographically, is Europe [3]. Therefore, it was only a matter of time before it was found in continental Europe, which our research group did in 2022 in urban/peri-urban populations of *R. norvegicus* and *R. rattus* in the city of Valencia (Spain) [4,5].

When *A. cantonensis* is found in its natural definitive hosts, to our knowledge, it is systematically reported in isolation, as if the rat lungworm were the only component of the within-rat parasite community, thus ignoring possible relationships with other parasite populations (infrapopulations) present in the infected rats. In this context, we report the first data on the concomitant parasite populations, i.e., protists, microsporidia and helminth coinfections, found in rats naturally infected by *A. cantonensis* in an urban rat population in the city of Valencia (Spain).

2. Materials and Methods

2.1. Study Area and Animals

Our research group signed an agreement with the Valencia City Council allowing us to investigate the presence of zoonotic parasites in rats trapped by the pest control company Laboratorios Lokímica, as part of the municipal pest control campaign in the city.

The presence of *A. cantonensis* was investigated in 125 rats, 97 *R. norvegicus* (43 males, 49 females and 5 indeterminate; 63 adults, 32 juveniles and 2 indeterminate) and 28 *R. rattus* (9 males and 19 females; 20 adults and 8 juveniles) trapped between April 2021 and March 2023. We studied rats captured at three trapping sites located in 17 of the 19 districts into which Valencia is divided, namely, in the sewer system (55 individuals), in city parks and gardens (43) and in orchards located in a peri-urban area of Valencia (27). Most of the rats, 74, were trapped in spring, 20 in autumn and 31 in winter. The trapped rats were kept at −20 °C until their parasitological examination.

Rat species were identified based on the external morphometry according to J. Gosàlbez [6]. Likewise, rats were considered juveniles or adults according to their body weight and external morphometry [6].

2.2. Parasitological and Molecular Techniques

Once thawed, the rats were dissected to extract the adult helminths from the different organs. The helminths found were studied by conventional helminthological techniques based on morphology. Cestodes and acantocephalans were stained with alcoholic hydrochloric carmine and trematodes with Grenacher's boracic carmine, for 24 h. Subsequently, these helminths were partially destained with acidified alcohol, dehydrated in an alcohol series, cleared with xylene and mounted in Canada balsam between slide and coverslip. Nematodes were, in turn, studied by direct examination between slide and coverslip with lactophenol as clearing fluid.

The nematodes found in the pulmonary arteries were also identified by molecular techniques [4,5]. The study of protozoans in the large intestine content was made by means of the Midi Parasep® SF (Apacor Ltd., Wokingham, UK) concentration technique followed by a multiplex PCR (AllplexTM Gastrointestinal Panel–Parasite Assay) for the detection of protist parasites such as *Giardia duodenalis, Entamoeba histolytica, Cryptosporidium* spp., *Blastocystis, Dientamoeba fragilis* and *Cyclospora cayetanensis* [7]. After DNA extraction, the microsporidia *Enterocytozoon bieneusi, Encephalitozoon intestinalis, E. cuniculi* and *E. hellem* were molecularly investigated by direct PCR [8]. *Toxoplasma gondii* was searched for in the rat brains by quantitative PCR (qPCR) [9] and the presence of *Leishmania infantum* in the spleens, ears and skin of the rats was also explored by qPCR [10].

2.3. Statistical Analysis

The comparison of the prevalences between intrinsic (age and sex) and extrinsic (site and season of capture) factors was made through the χ^2 test. Statistical significance was established at $p < 0.05$. Results obtained in both rat species, *R. norvegicus* and *R. rattus*, were analyzed and compared.

Statistical analysis was carried out using the IBM SPSS 26.0 for Windows (International Business Machines Corporation (IBM), Armonk, NY, USA) and StatView 5.0 (Statistical Analysis System (SAS) Institute Inc., Cary, NC, USA) software packages.

3. Results

3.1. *Angiostrongylus cantonensis Infected Rats*

The rat lungworm was identified in 14 of the 125 captured rats (11.20%), namely in 10 *R. norvegicus* (10.31%) and in 4 *R. rattus* (14.29%). It was found in the pulmonary arteries of 13 of the studied rats (10 *R. norvegicus* and 3 *R. rattus*). One black rat also harbored juvenile parasites in the brain and another one harbored the parasites exclusively in the brain. A total of 192 individuals of *A. cantonensis* were collected in the rats, with a mean intensity of 13.71 in the infected rats.

As for the sex of the rats, all *R. norvegicus* infected by *A. cantonensis* in the present study and two *R. rattus* (12/14 (85.71%)) were male. Only the two juvenile black rats were female (2/14, 14.29%)). The presence of *A. cantonensis* in *R. norvegicus* is sex-biased, with a higher prevalence of infection in males (10/43 (23.26%)) than in females (0/49); this difference is statistically significant ($\chi^2 = 12.785$, $p = 0.003$). This finding is supported by the fact that in *R. rattus*, in spite of the low number of black rats parasitized (4/28), and although males (2/9 (22.22%)) are more highly parasitized than females (2/19 (10.53%)), the difference between females of both rat species is also statistically significant ($\chi^2 = 5.314$, $p = 0.0212$).

Concerning the age of the rats, 11 rats infected by the nematode were adults (11/83 (13.25%)) and only 3 were juveniles (3/40 (7.5%)). No statistically significant differences were found concerning the age of both rat species together. However, considering each rat species separately, juvenile black rats have a higher prevalence (2/8 (25.00%)) than juveniles of the Norwegian rats (1/32 (3.13%)); this difference is statistically significant ($\chi^2 = 4.414$, $p = 0.0356$).

Angiostrongylus cantonensis was found in rats trapped in 7 of the 17 surveyed districts of Valencia (41.18%), as well as in the three trapping sites, i.e., in 6 rats of the 55 trapped in sewers (10.91%), in 2 of the 43 caught in parks (4.65%) and in 6 of the 27 captured in the orchards (22.22%) (Figure 1). Although the prevalences found are different according to the trapping sites, the results are not statistically significant when analyzed together due to the small sample sizes. However, there is a statistically significant difference between the prevalence found in parks compared to that found in orchards ($\chi^2 = 5.509$, $p = 0.0245$).

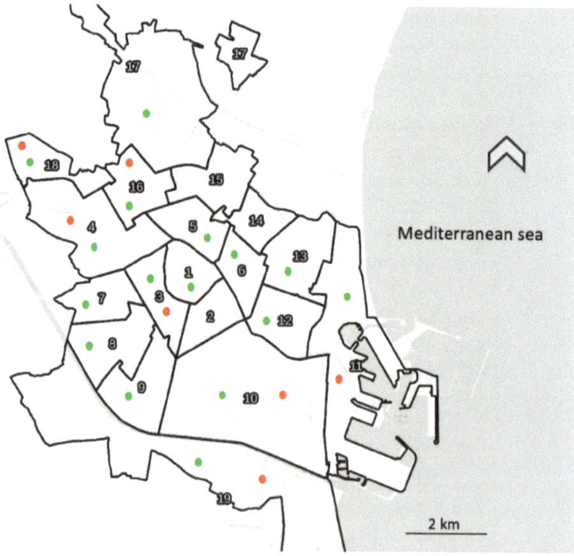

Figure 1. Map of the districts of Valencia; green dots showing the studied ones. *Angiostrongylus cantonensis*-infected rats were found in districts with red dots.

Of the 74 rats, 10 trapped in spring were infected by *A. cantonensis* (13.51%), while 2 were found parasitized in autumn (10%) and 2 in winter (6.45%).

3.2. Parasite Fauna/Coinfections

The parasite community in the studied organs of the 14 rats consisted of 19 different parasite species. Table 1 shows, in addition to *A. cantonensis*, the 18 other parasite species found according to rat species. Specifically, the parasite community of the rats studied consisted of six protists, one trematode, three cestodes, eight nematodes and one acanthocephalan. Fourteen of these nineteen species found are potentially zoonotic parasites, namely *Blastocystis*, *Giardia duodenalis*, *Cryptosporidium* spp., *Enterocytozoon bieneusi*, *Encephalitozoon hellem*, *Toxoplasma gondii*, *Brachylaima* spp., *Hydatigera taeniaeformis s.l.* larvae, *Hymenolepis nana*, *Hymenolepis diminuta*, *A. cantonensis*, *Calodium hepaticum*, *Gongylonema neoplasticum* and *Moniliformis moniliformis*. No rats were found infected by *Leishmania infantum*.

Table 1. Parasite species found in 10 *Rattus norvegicus* and 4 *Rattus rattus* in urban and peri-urban areas of Valencia, Spain, captured between 2021 and 2023.

Protists/Microsporidia Species	Microhabitat	Cycle	n (Host)	P (%)
Blastocystis	small intestine	M	8 (6 *Rn*, 2 *Rr*)	60 *Rn* 50 *Rr*
Giardia duodenalis	small intestine	M	9 (7 *Rn*, 2 *Rr*)	70 *Rn* 50 *Rr*
Cryptosporidium spp.	small intestine	M	1 (1 *Rn*)	10 *Rn*
Enterocytozoon bieneusi	small intestine	M	2 (1 *Rn*, 1 *Rr*)	10 *Rn* 25 *Rr*
Encephalitozoon hellem	small intestine	M	1 (1 *Rr*)	25 *Rr*
Toxoplasma gondii	brain	D	6 (5 *Rn*, 1 *Rr*)	50 *Rn* 25 *Rr*
Helminth species				
Brachylaima spp.	small intestine	H	1 (1 *Rn*)	10 *Rn*
Hydatigera taeniaeformis s.l. larvae	liver	H	5 (4 *Rn*, 1 *Rr*)	40 *Rn* 25 *Rr*
Hymenolepis nana	small intestine	H/M	3 (3 *Rn*)	30 *Rn*
Hymenolepis diminuta	small intestine	H	4 (3 *Rn*, 1 *Rr*)	30 *Rn* 25 *Rr*
Angiostrongylus cantonensis	pulmonary arteries/brain	H	14 [a] (10 *Rn*, 4 *Rr*)	100 *Rn* 100 *Rr*
Calodium hepaticum [b]	liver	M	11 (10 *Rn*, 1 *Rr*)	100 *Rn* 25 *Rr*
Mastophorus muris	stomach	H	2 (1 *Rn*, 1 *Rr*)	10 *Rn* 25 *Rr*
Eucoleus gastricus	stomach	M	3 (3 *Rn*)	30 *Rn*
Trichosomoides crassicauda	urinary bladder	M	3 (3 *Rn*)	30 *Rn*

Table 1. *Cont.*

Protists/Microsporidia Species	Microhabitat	Cycle	n (Host)	P (%)
Nippostrongylus brasiliensis	small intestine/ lungs (larvae)	M	9 (8 *Rn*, 1 *Rr*)	80 *Rn* 25 *Rr*
Heterakis spumosa	large intestine	M	4 (3 *Rn*, 1 *Rr*)	30 *Rn* 25 *Rr*
Gongylonema neoplasticum	esophagus/ stomach	H	2 (2 *Rn*)	20 *Rn*
Moniliformis moniliformis	small intestine	H	2 (2 *Rn*)	20 *Rn*

[a] In pulmonary arteries in 10 *Rn* and 3 *Rr*; in pulmonary arteries and brain in 1 *Rn* and only in brain in 1 *Rr*. [b] Range for *C. hepaticum* is not reported due to the difficulty in the reconstruction of dead parasites. Abbreviations: *Rn*, *Rattus norvegicus*; *Rr*, *Rattus rattus*; H, heteroxenous; M, monoxenous; n, number of parasitized hosts; P, prevalence. Potentially zoonotic species shaded in grey.

Table 2 summarizes the parasitic coinfections, i.e., the concomitant species found in the *A. cantonensis*-infected rats and their respective loads in the ten Norway rats as well as in the four black rats.

Table 2. Concomitant parasite species in naturally infected *Rattus norvegicus* (*Rr*) and *R. rattus* (*Rr*) by *Angiostrongylus cantonensis*: numbers represent helminth loads, and the total number of parasite species in individual rats is written in bold.

Protists/ Microsporidia Species	*Rn* I [s] /*Rr* * I [s]	*Rn* II [s] /*Rr* II [p]	*Rn* III [o] /*Rr* III [o]	*Rn* IV [o] /*Rr* * IV [o]	*Rn* V [o]	*Rn* VI [o]	*Rn* VII [p]	*Rn* VIII [s]	*Rn* IX [s]	*Rn* * X [s]
Blastocystis	+/+	+/−		−/+		+		+	+	+
G. duodenalis	+/+	+/+		+/−	+		+	+	+	
Cryptosporidium spp.									+	
T. gondii	+/−		−/+		+	+	+	+		
E. hellem	−/+									
E. bieneusi				−/+		+				
Helminth species										
Brachylaima spp.				4/−					4	
H. t. s.l. larvae	1/−	−/1	1/−					1		
H. nana								1	65	
H. diminuta		2/−	6/2	2/−		2	43			
A. cantonensis	2/30	4/9	7/3	13/2	7	3	33	4	19	57
C. hepaticum	+/−	+/−	+/+	+/−	+	+	+	+	+	+
M. muris			10/2			2				
E. gastricus				5/−			9		21	
T. crassicauda				4/−	4				14	
N. brasiliensis	80/−		65/2	31/−	6	2	25	HH	HH	
H. spumosa	9/−			2/7	1					
G. neoplasticum				1/−					3	
M. moniliformis		2/−							7	
Total n° species	**8**	**6**	**6**	**10**	**7**	**8**	**7**	**8**	**12**	**3**

s, Sewers; o, orchards; p, parks. *—juvenile. Abbreviations: *G. duodenalis*, Giardia duodenalis; *T. gondii*, Toxoplasma gondii; *E. hellem*, Encephalitozoon hellem; *E. bieneusi*, Enterocytozoon bieneusi; H. t. s.l. larvae, Hydatigera taeniaeformis s.l. larvae; *H. nana*, Hymenolepis nana; *H. diminuta*, Hymenolepis diminuta; *A. cantonensis*, Angiostrongylus cantonensis; *C. hepaticum*, Calodium hepaticum; *M. muris*, Mastophorus muris; *E. gastricus*, Eucoleus gastricus; *T. crassicauda*, Trichosomoides crassicauda; *N. brasiliensis*, Nippostrongylus brasiliensis; *H. spumosa*, Heterakis spumosa; *G. neoplasticum*, Gongylonema neoplasticum; *M. moniliformis*, Moniliformis moniliformis; HH, hundreds. + Numbers are not reported in *C. hepaticum* due to the difficulty in the reconstruction of dead parasites. Potentially zoonotic species shaded in grey.

Considering the coinfections in at least more than half of the 10 *A. cantonensis*-infected *R. norvegicus*, coinfection with *C. hepaticum* was the most frequent one, as all 10 rats infected by the rat lungworm were also infected by the hepatic nematode. Coinfection with *N. brasiliensis*—both species share the lungs as a microhabitat in their life cycles—was also frequent (in 8 out of 10 rats). No other helminth coinfection occurred in more than five rats, only concomitance with the protists, *Giardia* (in seven rats) and *Blastocystis* (in six rats).

More *R. rattus* infected with *A. cantonensis* would be needed to analyze coinfections in the case of the black rat.

4. Discussion

4.1. *Angiostrongylus cantonensis* Infected Rats

Published prevalences of *A. cantonensis* in rats can vary from 3 to 100% depending on the endemic area [11]. In our study, we found an overall prevalence of *A. cantonenesis* of 11.20% in Valencia, which—although not very high—was obtained in highly populated zones of the city, implying a potential (probably minimal) risk of acquiring the infection in those areas, mainly parks and gardens, in which infected snails could coexist with humans and, in particular, with children in playgrounds.

Regarding the sex of the rats, *A. cantonensis* was more prevalent in males of *R. norvegicus* than in females in the studied rat population. However, depending on the study, there was no difference found in the prevalence of the rat lungworm between males and females of the Norway rat, or *A. cantonensis* was found to be even more prevalent in females than in males [12,13]. Therefore, sex-biased parasitism seems to be a complex phenomenon influenced not only by hormones but also by other additional variables [12].

Concerning the influence of the trapping season, due to the small sample sizes, the results obtained are not statistically significant.

The fact that the highest *A. cantonenesis* prevalence was found in orchards is remarkable considering the high rate of consumption of raw vegetables in the Mediterranean diet, which poses a risk of acquiring the parasite larva through the ingestion of not-sufficiently washed salads [14].

Previously, we obtained a prevalence of the rat lungworm of 8.51% when 94 rats were studied [5]. The only data on the prevalence of the nematode near Spain was obtained in Tenerife, where a prevalence of 19.19% was obtained after studying 297 rats, most of them from rural areas [15]. Therefore, although the greater the number of rats studied, the higher the prevalence, there are no statistically significant differences between these figures.

4.2. Parasite Fauna/Coinfections

The studied rats presented a rich and varied within-host parasite community, the most remarkable finding being that 19 infrapopulations were found in only 14 rats, 14 of these parasite species being potentially zoonotic parasites posing a possible risk of transmission to the human population with which the rats coexist.

The *A. cantonensis*-infected rats captured in the sewers presented the greatest parasite species richness, as only the stomach nematode *Mastophorus muris* and the microsporidian *E. bieneusi* were not found (Table 2).

All components of the parasite community were found in rats trapped in spring (10/14). Only two rats captured in autumn and two in winter were found to be infected by *A. cantonensis*, so the absence of certain parasites in these individuals cannot be discussed.

Except for the microsporidian *Encephalitozoon hellem*, all other parasite species, 18, were found in the 10 *R. norvegicus* infected by *A. cantonensis*, and 12 different species were found in only 4 individuals of *R. rattus* (Table 1).

Only the 10 *A. cantonenesis*-infected *R. norvegicus* analyzed in this study presented a richer parasite community than the 100 Norway rat individuals we previously studied in Barcelona [7,10,16], without even considering the microsporidians and *T. gondii*, which were not investigated in the Barcelona rats.

Concerning coinfections, there was no case of monoparasitism among the studied rats. Adult *A. cantonensis*-infected *R. norvegicus* harbored from 6 to 12 different species in the same individual. The case of one adult *R. norvegicus* that harbored representatives of protists, trematodes, cestodes, nematodes as well as acanthocephalans in the intestine, with most of the helminths having high parasite loads (Table 2, *Rn* IX), is remarkable. Also noteworthy is the case of one juvenile Norway rat harboring 57 individuals of *A. cantonensis* (26 males and 31 females) in the pulmonary arteries as well as bearing a high burden of *C. hepaticum* infecting the liver (Table 2, *Rn* X). Both rats were trapped in the sewer system.

When analyzing the parasite community/coinfections found and the transmission routes, in the case of the monoxenous protists and microsporidia, rats became infected by the fecal/oral transmission route directly throughout the ingestion of cysts/oocysts/spores contaminating the environment, in particular the sewer system, and orchards, which are not normally irrigated with safe or potable water.

The presence of *T. gondii* in the rats may also be related to contamination by oocysts from cat feces or by cannibalism, a common occurrence in cases of limited food supply. The absence of the usual amount of food on the streets, due to the lockdown and the closure of restaurants during the pandemic, could have led to an increase in cannibalism that favored the *T. gondii* life cycle.

Considering the helminth parasites, 7 worms presented an indirect or heteroxenous life cycle and 5 had a monoxenous or direct cycle (Table 1). In the case of *H. nana*, the parasite is able to complete its life cycle either with the intervention of an arthropod intermediate host harboring the larval stage (cysticercoid) (heteroxenous life cycle) or without the intervention of any intermediate host but directly inside the intestine of the definitive host (monoxenous life cycle). Only one *R. norvegicus* presented a high *H. nana* load (*R.n.* IX in Table 2), which suggests the monoxenous-type cycle.

Among the monoxenous helminths, eggs shed in feces (or urine in the case of *Trichosomoides crassicauda*) are infective for the rats once the eggs embryonate in the soil. To become infected by *C. hepaticum*, a nematode that lives in the liver parenchyma, rats must also ingest the eggs that contaminate the environment. However, in this case, as the eggs are trapped in the liver, the rat must die in order to release the eggs, which mature in the soil. It is noteworthy that all the 10 *R. norvegicus* were infected by *C. hepaticum* (Table 2). This could indicate an increase in rat mortality during the pandemic period that ultimately favored cannibalism, leading to the release of eggs into the environment, enhancing the life cycle of *C. hepaticum*, as in the case of *T. gondii*.

In the case of *Nippostrongylus brasiliensis* (a murine model of *Necator americanus*), the larvae penetrate the skin of rats, or may also be ingested from the soil, and after molting and maturing in the lungs, reach the small intestine. The eggs are released in feces and hatch in the soil, releasing the L1 larvae, which become infective after molting. The nematode was found in almost all Norway rats at high burden levels (Table 2). Exceptionally, two of them (VIII and IX) harbored hundreds and hundreds of *N. brasiliensis*. It is hard to believe that this life cycle does not include processes of autoinfection and that the extraordinary number of adults in the intestine is due to repeated infections.

For heteroxenous life cycles, rats must ingest the eggs of *Hydatigera taeniaeformis* shed in cat (definitive host) feces. Rats act as intermediate hosts harboring the metacestode (strobylocercus) in the liver parenchyma. Cats are the only predator that rats have in cities, completing the biological cycle. Several rats harbored both *H. taeniaeformis* and *T. gondii*, parasites that share a common infection route, i.e., cat feces.

To become infected by *Brachylaima* spp. and *A. cantonensis*, rats must ingest infected snails (also slugs or paratenic hosts in the case of the rat lungworms). Two Norway rats were coinfected by both helminths, leading to the hypothesis that these two parasites could have shared a snail as an intermediate host.

For the rest of the heteroxenous helminths, intermediate hosts involve arthropods, mainly beetles for *H. diminuta* and cockroaches for *M. muris*, *G. neoplasticum* and *M. moniliformis*, with arthropods being an important element of the rat diet.

In terms of the host microhabitats for which the worms might compete, *A. cantonenesis* (adults, eggs and L1 larvae) and *N. brasiliensis* larvae (L3 and L4) share the same microhabitat, i.e., the lungs. In this regard, 8 of the 10 *R. norvegicus* and 1 *R. rattus* harbored both species (Table 2), so they do not appear to be competitors, at least in the studied rats. Likewise, six rats harbored *A. cantonensis* and *T. gondii*, parasites that share the brain as a microhabitat at a particular time of their life cycles.

The liver was also coinfected by the tapeworm larvae of *H. taeniaeformis* and *C. hepaticum* in three *R. norvegicus*, while the small intestine presented the greatest species richness, namely up to five different ones (Table 1). *Nippostrongylus brasiliensis* always occupies the first part of the small intestine, the duodenum, while the remaining helminths (*Hymenolepis* spp., *Brachylaima* spp. and *M. moniliformis*) are usually located in the jejunum and ileum.

Hosts which are ubiquitous, like rats, are more likely to become coinfected, as are hosts that occupy different ecological niches in which several parasites are present [17]. Consequently, rats that flourish in a wide range of environmental conditions, like sewers, parks, gardens and orchards in this particular case, are exposed to a greater diversity of parasites. Once infected, hosts use two strategies to cope with their parasites: resistance or tolerance [18]. Hosts can, by different mechanisms, reduce parasite burdens (resistance) or they can minimize the damage caused by the parasite load (tolerance). *Angiostrongylus cantonensis* was experimentally shown to cause a 10–20% mortality in *R. norvegicus* [19]. Also, an experimental study on parasite tolerance showed that mortality is related to the number of larvae of the rat lungworm used to infect rats [20]. However, it is difficult to know how to extrapolate these findings based on laboratory rats—not infected by any other parasite—to understand the consequences of coinfection in nature. In addition, coinfections can have negative effects on the host, accelerating its mortality or, otherwise, coinfections can have positive effects on the host, reducing its mortality [21].

5. Conclusions

Although we are aware of the limited number of rats studied, and considering that no histopathological studies were carried out to assess possible tissue damage, the total predominance of coinfected rats as well as their high parasite loads seem to indicate a trend towards parasite tolerance, at least in the studied rats infected by *A. cantonensis*. In this context, it seems clear that if at some point coinfections led to an increase in the mortality rate of the urban rat populations in Valencia, those populations that survived, considering their high reproductive capacity, may have given rise to tolerant populations that justify these high prevalences, parasite loads, and coinfections found in this study. In any case, further studies covering larger rat populations over the years as well as histopathological studies will help to determine whether tolerance is, in fact, the strategy that rat populations have developed against their parasites.

Author Contributions: Conceptualization, M.T.G.-P. and M.V.F.; methodology, M.T.G.-P., C.G., M.T., M.G.-S., J.S.-P., A.O., S.S.-D. and M.V.F.; formal analysis, M.T.G.-P. and M.V.F.; resources, M.T.G.-P., C.G., M.T., M.G.-S., J.S.-P., A.O., S.S.-D., R.B.-M. and M.V.F.; writing—original draft preparation, M.T.G.-P.; writing—review and editing, M.T.G.-P., M.T., M.G.-S., J.S.-P., A.O. and M.V.F. All authors have read and agreed to the published version of the manuscript.

Funding: This research received no external funding.

Institutional Review Board Statement: Ethics Committee approval was not necessary in our study, as all collections were conducted as part of the Municipal Pest Control Campaign. The use of snap-traps is one of the usual solutions for pest control. Following the Codes of Good Practice, these traps should be placed in areas or points not accessible to nontarget organisms (such as sewers or inside rodent safe stations in case of open areas) and checked frequently (every 4–5 days). We followed these criteria. There is no further legislation in our country on this issue.

Informed Consent Statement: Not applicable.

Data Availability Statement: No new data were created or analyzed in this study. Data sharing is not applicable to this article.

Acknowledgments: The authors would like to acknowledge the Health Service of the Valencia City Council for overseeing and promoting this research in the city.

Conflicts of Interest: The authors declare no conflicts of interest.

References

1. Cowie, R.H.; Ansdell, V.; Panosian Dunavan, C.; Rollins, R.L. Neuroangiostrongyliasis: Global Spread of an Emerging Disease. *Am. J. Trop. Med. Hyg.* **2022**, *107*, 1166–1172. [CrossRef] [PubMed]
2. Foronda, P.; López-González, M.; Miquel, J.; Torres, J.; Segovia, M.; Abreu-Acosta, N.; Casanova, J.C.; Valladares, B.; Mas-Coma, S.; Bargues, M.D.; et al. Finding of *Parastrongylus cantonensis* (Chen, 1935) in *Rattus rattus* in Tenerife, Canary Islands (Spain). *Acta Trop.* **2010**, *114*, 123–127. [CrossRef] [PubMed]
3. Paredes-Esquivel, C.; Sola, J.; Delgado-Serra, S.; Puig Riera, M.; Negre, N.; Miranda, M.A.; Jurado-Rivera, J.A. *Angiostrongylus cantonensis* in North African hedgehogs as vertebrate hosts, Mallorca, Spain, October 2018. *Eurosurveillance* **2019**, *24*, 1900489. [CrossRef] [PubMed]
4. Galán-Puchades, M.T.; Gómez-Samblás, M.; Osuna, A.; Sáez-Durán, S.; Bueno-Marí, R.; Fuentes, M.V. Autochthonous *Angiostrongylus cantonensis* Lungworms in Urban Rats, Valencia, Spain, 2021. *Emerg. Infect. Dis.* **2022**, *28*, 2564–2567. [CrossRef] [PubMed]
5. Galán-Puchades, M.T.; Gómez-Samblás, M.; Osuna, A.; Sáez-Durán, S.; Bueno-Marí, R.; Fuentes, M.V. Update on the first finding of rat lungworm, *Angiostrongylus cantonensis*, in *Rattus* spp. in continental Europe, Valencia, Spain, 2022. *Pathogens* **2023**, *12*, 567. [CrossRef] [PubMed]
6. Gosálbez, J. *Insectivors i Rosegadors de Catalunya. Metodologia d'Estudi i Cataleg Faunistic*; Ketres Editora: Barcelona, Spain, 1987; 241p.
7. Galán-Puchades, M.T.; Trelis, M.; Sáez-Durán, S.; Cifre, S.; Gosálvez, C.; Sanxis-Furió, J.; Pascual, J.; Bueno-Marí, R.; Franco, S.; Peracho, V.; et al. One Health Approach to Zoonotic Parasites: Molecular Detection of Intestinal Protozoans in an Urban Population of Norway Rats, *Rattus norvegicus*, in Barcelona, Spain. *Pathogens* **2021**, *10*, 311. [CrossRef] [PubMed]
8. Haro, M.; Izquierdo, F.; Henriques-Gil, N.; Andrés, I.; Alonso, F.; Fenoy, S.; del Aguila, C. First detection and genotyping of human-associated microsporidia in pigeons from urban parks. *Appl. Environ. Microbiol.* **2005**, *71*, 3153–3157. [CrossRef] [PubMed]
9. Reischl, U.; Bretagne, S.; Krüger, D.; Ernault, P.; Costa, J.M. Comparison of two DNA targets for the diagnosis of Toxoplasmosis by real-time PCR using fluorescence resonance energy transfer hybridization probes. *BMC Infect. Dis.* **2003**, *3*, 7. [CrossRef] [PubMed]
10. Galán-Puchades, M.T.; Gómez-Samblás, M.; Suárez-Morán, J.M.; Osuna, A.; Sanxis-Furió, J.; Pascual, J.; Bueno-Marí, R.; Franco, S.; Peracho, V.; Montalvo, T.; et al. Leishmaniasis in Norway Rats in Sewers, Barcelona, Spain. *Emerg. Infect. Dis.* **2019**, *25*, 1222–1224. [CrossRef] [PubMed]
11. Wang, Q.P.; Lai, D.H.; Zhu, X.Q.; Chen, X.G.; Lun, Z.R. Human angiostrongyliasis. *Lancet Infect. Dis.* **2008**, *8*, 621–630. [CrossRef] [PubMed]
12. Simões, R.O.; Júnior, A.M.; Olifiers, N.; García, J.S.; Bertolino, A.V.F.A.; Luque, J.L. A longitudinal study of *Angiostrongylus cantonensis* in an urban population of *Rattus norvegicus* in Brazil: The influences of seasonality and host features on the pattern of infection. *Parasites Vectors* **2014**, *7*, 100. [CrossRef] [PubMed]
13. Castillo, D.S.C.; Paller, V.G.V. Occurrence of *Angiostrongylus cantonensis* in rodents from the rice granary of the Philippines and associated risk factors for zoonotic transmission. *J. Parasit. Dis.* **2018**, *42*, 350–356. [CrossRef]
14. Prociv, P.; Spratt, D.M.; Carlisle, M.S. Neuroangistrongyliasis: Unresolved issues. *Int. J. Parasitol.* **2000**, *30*, 1295–1303. [CrossRef]
15. Martín-Carrillo, N.; Feliu, C.; Abreu-Acosta, N.; Izquierdo-Rodriguez, E.; Dorta-Guerra, R.; Miquel, J.; Abreu-Yanes, E.; Martin-Alonso, A.; García-Livia, K.; Quispe-Ricalde, M.A.; et al. A Peculiar Distribution of the Emerging Nematode *Angiostrongylus cantonensis* in the Canary Islands (Spain): Recent Introduction or Isolation Effect? *Animals* **2021**, *11*, 1267. [CrossRef]
16. Galán-Puchades, M.T.; Sanxis-Furió, J.; Pascual, J.; Bueno-Marí, R.; Franco, S.; Peracho, V.; Montalvo, T.; Fuentes, M.V. First survey on zoonotic helminthosis in urban brown rats (*Rattus norvegicus*) in Spain and associated public health considerations. *Vet. Parasitol.* **2018**, *259*, 49–52. [CrossRef] [PubMed]
17. Morand, S. Wormy world: Comparative tests of theoretical hypotheses on parasite species richness. In *Evolutionary Biology of Host-Parasite Relationships: Theory Meets Reality*; Poulin, R., Morand, S., Skorping, A., Eds.; Elsevier Science B.V.: Amsterdam, The Netherlands, 2000; pp. 63–79.
18. Rohr, J.R.; Raffel, T.R.; Hall, C.A. Developmental variation in resistance and tolerance in a multi-host–parasite system. *Funct. Ecol.* **2010**, *24*, 1110–1121. [CrossRef]
19. Garcia, J.S.; Dos Santos Bonfim, T.C.; Junior, A.M.; Tunholi, V.M.; Tunholi-Alves, V.M.; Mota, E.M.; Simoes, R.D.; Santana, A.C.; Hooper, C.; Pinheiro, J.; et al. Hematological and histopathological changes in *Rattus norvegicus* (Wistar) experimentally infected by *Angiostrongylus cantonensis* (Chen, 1935). *Parasitol. Int.* **2014**, *63*, 631–637. [CrossRef] [PubMed]

20. Ji, L.; Yiyue, X.; Xujin, H.; Minghui, Z.; Mengying, Z.; Yue, H.; Yanqi, W.; Langui, S.; Xin, Z.; Datao, L.; et al. Study on the tolerance and adaptation of rats to *Angiostrongylus cantonensis* infection. *Parasitol. Res.* **2017**, *116*, 1937–1945. [CrossRef]
21. Vaumourin, E.; Vourc'h, G.; Gasqui, P.; Vayssier-Taussat, M. The importance of multiparasitism: Examining the consequences of co-infections for human and animal health. *Parasites Vectors* **2015**, *8*, 545. [CrossRef]

pathogens

MDPI

Article

In the Dawn of an Early Invasion: No Genetic Diversity of *Angiostrongylus cantonensis* in Ecuador?

Luis Solórzano Álava [1], Cesar Bedoya Pilozo [1], Hilda Hernandez Alvarez [2], Lazara Rojas Rivera [2], Misladys Rodriguez Ortega [2], Jorge Fraga Nodarse [2], Leandro de Mattos Pereira [3], Raquel de Oliveira Simões [4] and Roberto do Val Vilela [5,*]

1 Hospital Luis Vernaza, Junta de Beneficencia de Guayaquil, Guayaquil 090101, Ecuador; lsolorzano@jbgye.org.ec (L.S.Á.); cbedoya@jbgye.org.ec (C.B.P.)

2 Instituto de Medicina Tropical Pedro Kouri, La Habana 17100, Cuba; hilda@ipk.sld.cu (H.H.A.); lrojas@ipk.sld.cu (L.R.R.); misladys@ipk.sld.cu (M.R.O.); fraga@ipk.sld.cu (J.F.N.)

3 Instituto Tecnológico Vale, Belém 66055-090, PA, Brazil; leandro.mattos@pq.itv.org

4 Departamento de Parasitologia Animal, Instituto de Veterinária, Universidade Federal Rural do Rio de Janeiro, Seropédica 23890-000, RJ, Brazil; raquelsimoes@ufrrj.br

5 Instituto Oswaldo Cruz, Fundação Oswaldo Cruz, Rio de Janeiro 21040-900, RJ, Brazil

* Correspondence: roberto.vilela@ioc.fiocruz.br; Tel.: +55-(21)-2562-1011

Abstract: The nematode *Angiostrongylus cantonensis* has been reported worldwide. However, some basic questions remain unanswered about *A. cantonensis* in Ecuador: (1) Was the invasion of *A. cantonensis* in Ecuador unique, or did it occur in different waves? (2) Was this invasion as recent as historical records suggest? (3) Did this invasion come from other regions of South America or elsewhere? To address these issues, we assessed the genetic diversity of MT-CO1 gene sequences from isolates obtained in 11 of Ecuador's 24 provinces. Our Bayesian inference phylogenetic tree recovered *A. cantonensis* as a well-supported monophyletic group. All 11 sequences from Ecuador were identical and identified as AC17a. The haplotype AC17a, found in Ecuador and the USA, formed a cluster with AC17b (USA), AC13 (Thailand), and AC12a-b (Cambodia). Notably, all the samples obtained in Ecuadorian provinces' different geographic and climatic regions had no genetic difference. Despite the lack of genetic information on *A. cantonensis* in Latin America, except in Brazil, our finding differs from previous studies by its absence of gene diversity in Ecuador. We concluded that the invasion of *A. cantonensis* in Ecuador may have occurred: (1) as a one-time event, (2) recently, and (3) from Asia via the USA. Further research should include samples from countries neighboring Ecuador to delve deeper into this.

Keywords: Invasive species; eosinophilic meningoencephalitis; cytochrome c oxidase subunit I

check for updates

Citation: Solórzano Álava, L.; Bedoya Pilozo, C.; Hernandez Alvarez, H.; Rojas Rivera, L.; Rodriguez Ortega, M.; Fraga Nodarse, J.; Pereira, L.d.M.; Simões, R.d.O.; Vilela, R.d.V. In the Dawn of an Early Invasion: No Genetic Diversity of *Angiostrongylus cantonensis* in Ecuador?. *Pathogens* **2023**, *12*, 878. https://doi.org/10.3390/pathogens12070878

Academic Editor: Susan Jarvi

Received: 31 March 2023
Revised: 30 April 2023
Accepted: 1 May 2023
Published: 27 June 2023

1. Introduction

The rat lungworm *Angiostrongylus cantonensis* (Chen, 1935) was first described in the bronchi of the rodents *Rattus rattus* (Linnaeus, 1758) and *Rattus norvegicus* (Berkenhout, 1769) in Guangzhou (formerly Canton), China [1]. This nematode is the etiological agent of neuroangiostrongyliasis, which is the leading cause of eosinophilic meningitis (EM) or eosinophilic meningoencephalitis (EME) in humans, an infectious disease of the central nervous system [2]. This disease is characterized primarily by increased eosinophils in peripheral blood and cerebrospinal fluid, among other symptoms such as fever and severe headache [3,4].

The first documented human case of neuroangiostrongyliasis was in Taiwan in 1944. However, it took nearly two decades to establish a clear link between the parasite and the disease (i.e., *A. cantonensis* as a causative agent of EME) [5]. Since the first report, several outbreaks have been reported globally as the parasite has spread from traditional endemic regions of Southern China and Southeast Asia to the Pacific islands, Japan, Australia,

Africa, the Canary Islands, the Balearic Islands, and the Americas, including the USA, Caribbean islands, and Brazil [2,4,6,7]. In Europe, *A. cantonensis* infections have been reported in different countries, although apparently, only one case was autochthonous [8]. By 2008, more than 2800 cases of human angiostrongyliasis had already been recorded in 30 countries [9]. The spread of parasites in different regions threatens people living in endemic areas and a growing number of travelers visiting these regions [10].

In 2008, *A. cantonensis* was reported for the first time in Ecuador parasitizing the giant African land snail *Achatina* (*Lissachatina*) *fulica* Bowdich, 1822 and the rats *R. rattus* and *R. norvegicus* [11,12]. Since then, outbreaks and isolated cases have been reported to the Ministry of Public Health of Ecuador (MSP) [13], with most clinical-epidemiological suspicion and one necropsy-confirmed case [14]. The parasite is now considered endemic throughout most of the country [11,15]. The invasive pest *A. fulica* is one of the main intermediate hosts for *A. cantonensis* [16]. This mollusk lives in urban and rural areas and plays a vital role in the spread of the parasite [17]. Humans may become infected by ingesting food contaminated with third-stage larvae or eating infected raw snails [2]. Thus, *A. fulica* is an essential transmitter of eosinophilic meningoencephalitis and ocular angiostrongyliasis [18].

Different molecular biology methods have been employed to detect *A. cantonensis* [19–24]. Furthermore, they have been applied to explore systematic and population genetic aspects of *Angiostrongylus* taxa since populations have significant variability [25–32]. The use of mitochondrial genes, such as cytochrome c oxidase subunit I (MT-CO1), as molecular markers for specific parasite identification has been efficient [33–36]. The MT-CO1 gene has been used in studies on phylogeny, phylogeography, and haplotype identification [37–40]. However, some basic questions remain unanswered about *A. cantonensis* in Ecuador: (1) Was the invasion of *A. cantonensis* in Ecuador a single event, or did it occur in different waves? (2) Was this invasion as recent as historical records suggest? (3) Did this invasion come from other regions of South America or elsewhere?

To tackle these questions, we assessed the genetic diversity of MT-CO1 gene sequences from isolates obtained in 11 of Ecuador's 24 provinces. Thus, we verified how many lineages could be found in different regions of Ecuador and whether there was enough time for the lineages to diversify. We also established these isolates' phylogenetic and phylogeographic relationships, comparing them with other sequences from South America and the rest of the world. Consequently, we could retrace the possible origin of the lineages found in Ecuador.

2. Materials and Methods

2.1. Parasites and Experimental Infection

Third-stage larvae (L_3) were obtained from *A. fulica*, collected in 11 provinces of Ecuador (Figure 1) using the catch-per-unit-effort method for 30 min in each locality [16]. The L_3 were used to experimentally infect 12-week-old adult female Wistar strain *R. norvegicus* rats (200 ± 2 g body mass). The Instituto de Investigación en Salud Pública (INSPI) vivarium supplied the rats with their corresponding health and genetic quality certificates. The cycle was maintained in the National Reference Center for Parasitology. An average of 150 L_3 (counted in a Neubauer chamber) were orally administered to each rat using a pipette. Infected rats were separated into cages (two specimens per cage) and identified according to the locality (province) where the infected gastropods were collected. Rats were kept under controlled temperature (21–24 °C) and humidity (60%), alternating 12-h light/dark cycle, and received food and water at pH 7.0. All procedures followed the guidelines for the maintenance and use of laboratory animals, following the specific legislation covering animals used for scientific purposes Directive 2010/63/EU as amended by European Union (EU) Regulation 2019/1010 [41].

Figure 1. Map of Ecuador showing the study area highlighting the sampled provinces (https://provinciasecuador.com/mapa-politico-del-ecuador/ (accessed on 30 April 2023) with modifications).

Thirty-five days after administration of the larvae, the rats were euthanized using CO_2. The thoracic cavity (heart, pulmonary arteries, and lungs) was examined for parasitic nematodes (juvenile or adult) according to protocols previously established at INSPI [11]. Specific taxonomic characteristics such as caudal bursa and the spicule length were used to identify the nematodes [42,43]. Approximately 30–40 adult *A. cantonensis* specimens from two infected rats representing each province were stored in a sterile labeled 50 mL Falcon tube with 70% ethanol in an ultra freezer at −80 °C.

2.2. Molecular Phylogenetic and Phylogeographic Analyses

We used DNA sequences obtained from adult parasites, as previously reported [39,44,45], to conduct phylogenetic and phylogeographic studies. Genomic DNA samples were isolated from adult parasites recovered from the rats representing each province. Before DNA isolation, the nematodes were partitioned into tiny pieces with a scalpel and suspended in saline (0.9% NaCl). We used the QIAamp DNA Mini Kit (QUIAGEN, Hilden, Germany) for DNA isolation according to the manufacturer's protocol. Each isolated DNA sample was identified according to its origin and stored at −80 °C until further amplification by PCR technique. Genomic DNA concentration was measured directly in a NanoDrop 2000 spectrophotometer (Thermo Fisher Scientific, Walthan, MA, USA).

DNA isolated from approximately 30 adult parasites was subjected to PCR to amplify the mitochondrially encoded cytochrome c oxidase I (MT-CO1) gene [37]. PCR reactions were performed in a 25 µL total volume containing 12.5 µL of GoTaq Colorless Master Mix (Promega, Madison, WI, USA: 2× DNA polymerase, 400 µM dATP, 400 µM dGTP, 400 µM dCTP, 400 µM dTTP, and 3 mM MgCl2, pH 8.5); 1.5 µL of 10 µM each MT-CO1 gene primer (Thermo Fisher Scientific, Walthan, MA, USA); 5.5 µL of distilled water; and 4 µL of genomic DNA. We also used a positive control consisting of an adult parasite DNA obtained from a wild-type rat (*R. rattus*) and a negative control with ultrapure water. The primers used were:

co1-F (5′TAAAGAAAGAAAGAACATAATGAAAATG3′)

co1-R (3′TTTTTTTTTGGCATTCCTGAGGAGGT5′)

Modifications have been made to the original thermal cycling protocol by Vitta et al. [37] to standardize the technique in the INSPI laboratory and obtain the desired amplicons of approximately 450 base pairs (bp) as follows: 94 °C for 5 min; followed by 30 cycles of 94 °C for 1 min, 48 °C for 30 s, and 72 °C for 60 s; with a final extension at 72 °C for 5 min. PCR was performed in a C1000 Touch thermal cycler (Bio-Rad, Hercules, CA, USA).

We verified PCR products after 1.2% agarose (Promega, Madison, WI, USA) gel electrophoresis in 0.04 M Tris-acetate running buffer, 1 Mm ethylenediamine tetraacetic acid, pH 8.0 (Thermo Fisher Scientific, Walthan, MA, USA). We added 10 µL of Syber® 1× (10,000×) dye (Thermo Fisher Scientific, Walthan, MA, USA) to the agarose gel. We added Blue/Orange Loading Dye, 6× (Promega, Madison, WI, USA), to each sample. TrackIt 100 bp DNA Ladder (0.1 µg/µL), with 100 to 1000 bp range (Thermo Fisher Scientific, Walthan, MA, USA), was used as molecular weight marker. Electrophoresis was performed at 80 V for 55 min using a PowerPac HC power supply (Bio-Rad, Hercules, CA, USA). PCR products were visualized using the ChemiDoc XRS imaging system (Bio-Rad, Hercules, CA, USA).

Amplicons purification; cycle-sequencing of both strands via the Sanger method, using the abovementioned PCR primers; and product precipitation, formamide resuspension, and analysis using the 3130 DNA Analyzer (Applied Biosystems, Foster City, CA, USA) were performed at the biochemistry department of the Universidad de las Américas (Ecuador).

The resulting chromatograms were edited with the software platform Geneious R7.0 (Biomatters, Aukland, New Zealand) [44]. Sense and anti-sense sequences of each amplified and sequenced sample were assembled into contigs. The resulting consensus sequences corresponding to 11 Ecuadorian provinces were deposited in the GenBank (Table 1).

Table 1. Identification and GenBank accession numbers of sequences obtained in this study, followed by their respective sampling localities.

Identification	GenBank Accession Number	Province
LSA-01	MW391020	Esmeraldas
LSA-02	MW390970	Santa Elena
LSA-03	MW390971	El Oro
LSA-04	MW390972	Guayas
LSA-05	MW390967	Zamora
LSA-06	MW390974	Pastaza
LSA-07	MW390969	Orellana
LSA-08	MW390973	Manabi
LSA-09	MW390968	Napo
LSA-10	MW390966	Los Rios
LSA-11	MW390965	Sucumbios

To construct our MT-CO1 dataset, we used *A. cantonensis* sequences found in GenBank that overlapped ours (Table S1). As outgroups, we added one sequence of *Angiostrongylus mackerrasae* Bhaibulaya, 1968 (MN793157) and three sequences of *Angiostrongylus malaysiensis* Bhaibulaya and Cross, 1971 (KT947979, KU532150, KU532153), all from GenBank (Table S1). Sequences in the dataset were aligned by multiple alignments using

MUSCLE [45], under default parameters, within the Geneious package. Final manual trimming of non-overlapping regions of the alignment was carried out using the Mesquite 3.70 software package [46].

Two different matrices were used in this study. In the first matrix, used for phylogenetic inferences, we excluded all duplicated sequences, keeping only one copy of each haplotype of *A. cantonensis* and the outgroup. In the second matrix, used for phylogeographic analyses, we included all *A. cantonensis* sequences and excluded the outgroup. To find the optimal partition clustering arrangements and corresponding log(ml) values in both matrices, we conducted Bayesian clustering of linked molecular data analyses using BAPS 6.0 [47,48].

Bayesian inference (BI) phylogenetic analyses were conducted using MrBayes 3.2.6 [49] on XSEDE within the CIPRES Science Gateway [50]. We used independent GTR+I+G models for each codon position, with unlinking of base frequencies and parameters. Sampling was performed by MCMC, for 10,000,000 generations, with four simultaneous chains, in two runs. Node supports were given by Bayesian posterior probabilities (BPP) of trees sampled every 100 generations after removal of the first 25% 'burn-in' generations. We assessed sampling adequacy using the program Tracer 1.7.1 [51] to calculate the effective sample sizes (ESSs) of parameters. We considered robust values above 200 effectively independent samples.

An intraspecific phylogeographic network was inferred using the program PopART, version 1.7 [52], with the median-joining network method [53]. Using DnaSP 6.12.03 [54], we organized the sequences into groups according to their geographic localities (countries). We also calculated, using DnaSP, the genetic diversity by the numbers of haplotypes (H), polymorphic sites (S), haplotype diversity (Hd), and nucleotide diversity (π). We finally used DnaSP for neutrality tests Tajima's D [55] and Fu's Fs [56].

3. Results

Along with our 11 MT-CO1 gene sequences of *A. cantonensis* from Ecuador, we added 105 sequences of *A. cantonensis* from GenBank and four sequences of outgroups. The full dataset had 120 sequences of *Angiostrongylus* ranging from 255 to 1617 bp in length (Table S1). The haplotypes were named AC1-17, following the names for haplotypes previously adopted [38,40,57], adding letters to variants. All 11 sequences from Ecuador were identical and identified as AC17a. The Ecuador sequences were identical to five sequences from New Orleans, Louisiana, USA (USA-LA), retrieved from GenBank.

3.1. Molecular Phylogenetic Analyses

After multiple sequence alignment, trimming, and removal of all duplicates in the first matrix for phylogenetic inferences, the matrix resulted in 29 taxa and 255 sites. Of these, 201 were constant characters, and 41 were variable parsimony-informative characters. *Angiostrongylus cantonensis* was represented by 25 sequences, while the outgroup by four. According to the population structure recovered using BAPS, *Angiostrongylus* specimens were distributed in five clusters in the 29 sequences matrix.

After 25% burn-in removal, the BI mean estimated marginal likelihood was −751.4969, and the median was −751.1709. The ESS values were well above 200 for all parameters. The BI phylogenetic tree (Figure 2) recovered *A. cantonensis* as a well-supported monophyletic group (BPP = 1.00). Within *A. cantonensis*, the sequence AC17a, from Ecuador and USA-LA, was in a polytomy with AC17b (USA-LA); AC5a (Brazil, Japan, French Polynesia, and Hawaii, USA); AC5b (Japan); AC13 (Thailand); and a moderately supported clade (BPP = 0.70), formed by sequences AC8a (Brazil) and AC8b (Australia, Balearics, Canaries, Taiwan, and USA-LA). This polytomy was moderately supported (BPP = 0.78) and formed another polytomy with sequences AC12a and AC12b from Cambodia. This more inclusive polytomy was strongly supported (BPP = 0.98) and coincided with Cluster 3.

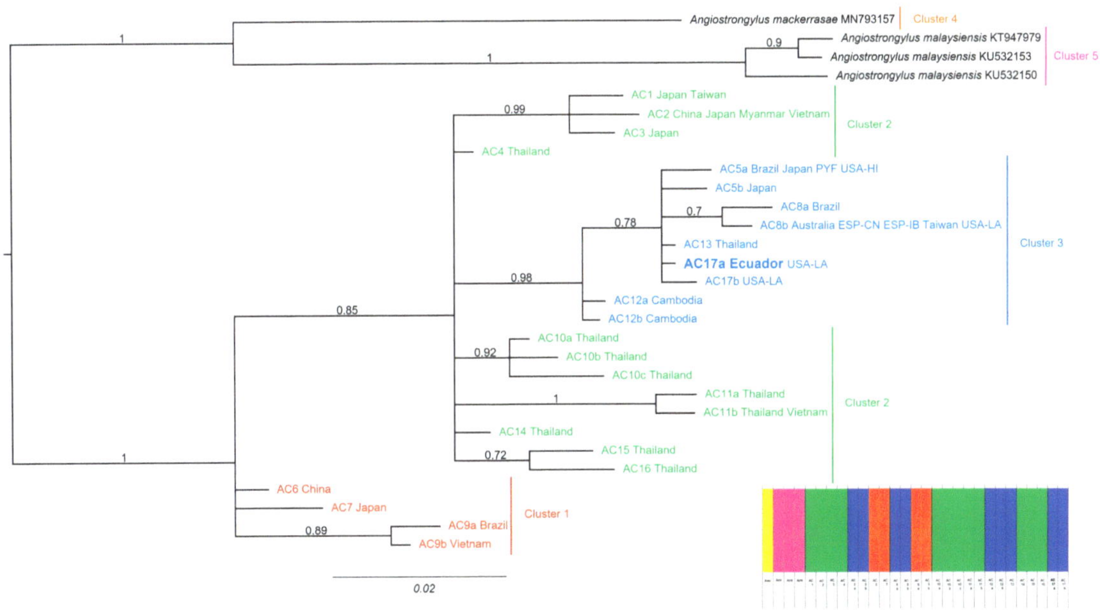

Figure 2. Bayesian inference (BI) phylogenetic relationships of *A. cantonensis* specimens and outgroups unique MT-CO1 gene sequences (255 bp). GenBank accession numbers of *A. cantonensis* sequences are provided in Supplementary Material Table S1. The values at the nodes are BPPs (>0.50). The scale bar is the number of substitutions per site. Sequence labels are colored based on the clusters recovered in the BAPS cluster analysis (bottom right). Sequences are labeled AC1-17, following names for haplotypes previously adopted [38,40,57], adding letters to variants, followed by the localities (countries) where they are found. Clusters 1–5 were recovered in the BAPS cluster analysis for the 29 sequences matrix.

3.2. Phylogeographic Analyses

The second matrix, for phylogeographic analyses, included only sequences of *A. cantonensis*. This dataset included 11 sequences from Ecuador and 105 sequences from GenBank, excluding the outgroup, totaling 116 taxa and 255 sites after multiple sequence alignment and trimming. The total number of sites, excluding sites with gaps or missing data, was 254. The number of haplotypes was H = 25, the number of polymorphic sites S = 36, the haplotype diversity Hd = 0.895, the nucleotide diversity $\pi = 0.02546$, Fu's Fs = -2.380, and Tajima's D = -0.42728 ($p > 0.10$).

According to the population structure recovered using BAPS, *A. cantonensis* specimens were distributed in seven clusters in the 116 sequences matrix. We indicated the clusters in the intraspecific phylogeographic network (Figure 3). The haplotype AC17a, from Ecuador and USA-LA, formed a cluster with AC17b (USA-LA), AC13 (Thailand), and AC12a-b (Cambodia). This haplogroup was labeled Cluster 5 in the network.

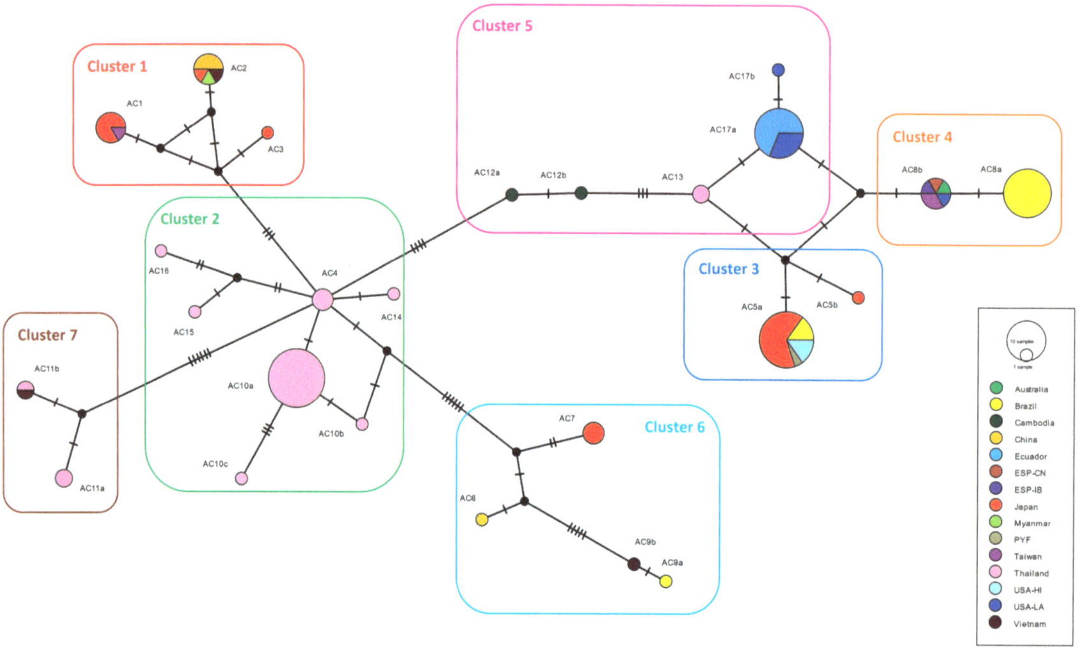

Figure 3. Median-joining haplotype network of *A. cantonensis* (25 haplotypes) based on 116 partial MT-CO1 gene sequences (255 bp). The size of the circles represents the frequency of haplotypes. The colors of the circles represent the localities (countries) of occurrence of each haplotype. Black circles are median vectors. Sequences are labeled AC1-17, following names for haplotypes previously adopted [38,40,57], adding letters to variants. Clusters 1–7 were recovered in the BAPS cluster analysis for the 116 sequences matrix.

4. Discussion

Introducing non-native mollusks, such as *A. fulica*, is essential in transmitting *A. cantonensis* [58]. Since the mid-20th century, *A. fulica* has been introduced into the tropics and subtropics and is considered the most harmful snail pest in these regions [17]. In Brazil, these mollusks were possibly introduced more than once on different occasions [59]. The first account is from the mid-1970s in the state of Minas Gerais [60]. The second, better documented, and probably the chief introduction was in the late 1980s in the state of Paraná from specimens brought from Indonesia for commercial purposes (snail farming) that were unsuccessful [61]. The giant African snail is now widespread in all 26 Brazilian states and the Federal District [62,63].

According to data from an Ecuadorian government organization, these snails were brought into the country in the mid-1990s. As in Brazil, this was for commercial purposes. Snail farms were built in some valleys of the Ecuadorian highlands, which offered an ideal temperature between 17 °C and 25 °C [64]. However, their breeding did not provide the expected economic returns. Inevitably, most of the farms were abandoned, and the snails were released into the environment. The result was a widespread infestation of urban and rural areas in almost all of Ecuador's provinces [12]. *Achatina fulica* was probably the vector that introduced *A. cantonensis* to the country, as in Brazil [65] and China [66].

As for the definitive hosts, it is presumed that *R. rattus* arrived in Ecuador between the 16th and 17th centuries with the ships of the Spanish conquistadors [67]. *Rattus norvegicus* probably originated in China and spread to Europe, reaching North America through shipping during the second half of the 18th century. Both species are now widely distributed in urban areas worldwide [68].

The existence of intermediate and definitive hosts in almost all of Ecuador has contributed to the endemic nature of angiostrongyliasis distribution, making the control of this disease even more complex [11]. In 2008, the snail *A. fulica* (intermediate host) and the rat *R. rattus* (definitive host) were found naturally infected by *A. cantonensis* in Ecuador [11,12]. Both intermediate (*A. fulica*) and definitive (*R. rattus*) hosts are non-native species to Ecuador and are considered among the 100 most important invasive species in the world, according to the World Conservation Union [69]. Invasive species in an ecosystem can affect biotic alter interactions, impacting the economy, the environment, or public and animal health [70,71]. Moreover, elder Ecuadorians' habit of eating raw snails increases the risk of *A. cantonensis* infection [12].

Earlier studies using the MT-CO1 to distinguish *A. cantonensis* isolates have shown different geographical isolates in determinate regions [38–40,57]. Tokiwa et al. [40] reported seven different haplotypes (AC1 to AC7): five were found in Japan (AC1, AC2, AC3, AC5, and AC7), two in mainland China (AC2 and AC6) and only one in Taiwan (AC1). In Brazil, analyses from 15 geographic isolates determined the presence of three different MT-CO1 haplotypes (AC5, AC8, and AC9). Most sample sequences were AC5 or AC8, whereas AC9 was a new haplotype [38]. Rodpai et al. [57] identified different *A. cantonensis* haplotypes in Cambodia, Myanmar, Thailand, and Hawaii, USA. Two of them (AC2 and AC5) had been previously reported. The AC2 haplotype, previously reported in China and Japan, was found in Myanmar. The AC5 haplotype, previously reported in Brazil and Japan, was found in Hawaii. Four new haplotypes (AC10-AC13) were also reported in Southeast Asia [57].

Such studies have shown that *A. cantonensis* in Asia has greater genetic diversity [39,40,57], indicating that this parasite has been circulating in these regions for a long time. Conversely, the sequence diversity of *A. cantonensis* is lower in many areas outside Asia [72]. Otherwise, there is little or no genetic information on the parasite in other regions of the planet, such as the Americas, except in Brazil [38].

In the present study, all sequences of the isolates from Ecuador were identical, the haplotype AC17a. In our phylogenetic analyses, this haplotype was nested into a polytomy with other sequences from around the world. Remarkably, all samples were obtained from provinces of Ecuador in different geographic and climatic regions, yet they did not show any genetic divergence between them.

The findings reported here represent a novelty in studying the genetic diversity of *A. cantonensis* isolates. Although there is a need for more information on the genetic diversity of this parasite in other Latin American countries, except for Brazil, our results are different from previous studies due to the complete absence of genetic diversity of *A. cantonensis* in Ecuador. Even admitting that the low number of nucleotide base pairs obtained could make the sequence homogeneous in the isolates from Ecuador, this same region of the MT-CO1 gene showed variations in the other haplotypes compared.

The fact that only one haplotype was found in 11 different Ecuadorian provinces is revealing. It strongly advocates a single introduction event. Furthermore, this result suggests that *A. cantonensis* has been recently introduced in the country, as there was no time for new haplotypes to differentiate from the original. This may justify the non-existence of genetic diversity among different circulating isolates.

Interestingly, the sequences from Ecuador shared a recent common ancestor with two Brazilian haplotypes (AC5 and AC8) [73]. However, it is unlikely that this could indicate a historical connection between the strains from both countries. The AC5 haplotypes found in Brazil from Pirituba (state of São Paulo), Queimados, and Niterói (state of Rio de Janeiro) correspond to a haplotype found in Japan, Hawaii, and French Polynesia [57,72,74], suggesting that the arrival of the parasite in Rio de Janeiro or São Paulo may have occurred from the Asian continent [38]. This hypothesis is also considered for the AC8a haplotype, closely related to AC8b, found in Australia, the Balearics, the Canaries, Taiwan, and the United States of America (USA). This shows the possible spread of *A. cantonensis*, with the giant African land snail, as a vector, from the arrival localities in Brazil to the Southeast, Northeast, and North Brazilian regions [38].

The sequences obtained here also grouped with AC13 and AC17 haplotypes from Thailand and the USA, respectively. The haplotypes AC10, AC11, and AC13, from Thailand, and AC12, from Cambodia, were described by Rodpai et al. [57] in phylogenetic studies using different DNA regions of *A. cantonensis* and *A. malaysiensis*. The haplotypes AC17, from the USA, were reported in a survey to identify *A. cantonensis* and determine the association between ecological characteristics and factors related to definitive hosts (*R. rattus*, *R. norvegicus*, *Sigmodon hispidus*, and *Oryzomys palustris*) associated with transmission risk of angiostrongyliasis in New Orleans [75]. The haplotypes AC12, AC13, and AC17 formed a cluster in the haplotype cluster analysis, suggesting that *A. cantonensis* may have arrived in Ecuador from Asia via the USA.

5. Conclusions

Our results suggest that the invasion of *A. cantonensis* in Ecuador occurred as a single event since only one haplotype was present in all 11 provinces studied, encompassing different ecoregions of Ecuador. Moreover, this invasion may have occurred recently, as we found no variation from the initial haplotype. It is unlikely that *A. cantonensis* reached Ecuador from Brazil. It is conceivable that the lineage found in Ecuador came from Asia via the USA. Future studies should sample countries neighboring Ecuador to infer migratory routes into this country in more detail.

Supplementary Materials: The following supporting information can be downloaded at: https://www.mdpi.com/article/10.3390/pathogens12070878/s1, Table S1: 120 documents from Angiostrongylus.txt.

Author Contributions: Conceptualization, L.S.Á., L.R.R. and H.H.A.; methodology, H.H.A., L.R.R., L.S.Á. and R.d.V.V.; software, C.B.P., L.d.M.P. and R.d.V.V.; validation, J.F.N. and M.R.O.; formal analysis, C.B.P., J.F.N., L.d.M.P. and R.d.V.V.; investigation, L.S.Á. and H.H.A.; resources, L.S.Á. and R.d.V.V.; data curation, C.B.P., L.S.Á. and R.d.V.V.; writing—original draft preparation, C.B.P. and L.S.Á.; writing—review and editing, C.B.P., L.S.Á., R.d.V.V. and R.d.O.S.; visualization, R.d.V.V. and R.d.O.S.; supervision, L.S.Á.; project administration, L.S.Á.; funding acquisition, L.S.Á. All authors have read and agreed to the published version of the manuscript.

Funding: This research received no external funding.

Institutional Review Board Statement: All procedures followed the guidelines for the maintenance and use of laboratory animals, following the specific legislation covering animals used for scientific purposes Directive 2010/63/EU as amended by European Union (EU) Regulation 2019/1010.

Informed Consent Statement: The protocol was approved by the Human Research Ethics Commission (CEISH) of the Kennedy Clinic of the City of Guayaquil, supported by the Ministry of Health of Ecuador, unique approval code DIS-CEISH-HCK-09-014. Doctors were completely free to decide to participate in the investigation. At the end of the survey, they signed and entered the number of their identification document in it.

Data Availability Statement: The data presented in this study are openly available in GenBank, accession numbers MW390965 to MW390974 and MW391020.

Acknowledgments: The authors thank members of the Parasitology Laboratory, Centro de Referencia Nacional de Parasitología y Micología, Instituto de Investigación en Salud Pública, INSPI, and the staff of the Microbiology laboratory, Facultad de Medicina, Universidad de Guayaquil. We are also grateful to Alessandra L. Morassutti and Arnaldo Maldonado Júnior, who put us in contact to carry out this work.

Conflicts of Interest: The funders had no role in the study's design; in the collection, analyses, or interpretation of data; in the writing of the manuscript; or in the decision to publish the results. The researchers declare that they have no conflict of interest.

References

1. Chen, H.-T. Un Nouveau Nématode Pulmonaire, Pulmonema Cantonensis, n. g., n. Sp. *Ann. Parasitol. Hum. Comp.* **1935**, *13*, 312–317. [CrossRef]
2. Eamsobhana, P. Eosinophilic Meningitis Caused by Angiostrongylus Cantonensis—A Neglected Disease with Escalating Importance. *Trop. Biomed.* **2014**, *31*, 569–578. [PubMed]
3. Martins, Y.C.; Tanowitz, H.B.; Kazacos, K.R. Central Nervous System Manifestations of Angiostrongylus Cantonensis Infection. *Acta Trop.* **2015**, *141*, 46–53. [CrossRef]
4. Wang, Q.P.; Wu, Z.D.; Wei, J.; Owen, R.L.; Lun, Z.R. Human Angiostrongylus Cantonensis: An Update. *Eur. J. Clin. Microbiol. Infect. Dis.* **2012**, *31*, 389–395. [CrossRef] [PubMed]
5. Alicata, J.E. Biology and Distribution of the Rat Lungworm, Angiostrongylus Cantonensis, and Its Relationship to Eosinophilic Meningoencephalitis and Other Neurological Disorders of Man and Animals**Paper Presented at the First International Congress of Parasitology, Rome, Italy, 19–26 September 1964. In *Advances in Parasitology*; Dawes, B., Ed.; Academic Press Inc.: London, UK, 1965; Volume 3, pp. 223–248.
6. Maldonado-Júnior, A.; Simões, R.O.; Oliveira, A.P.M.; Motta, E.M.; Fernandez, M.A.; Pereira, Z.M.; Monteiro, S.S.; Torres, E.J.L.; Thiengo, S.C. First Report of Angiostrongylus Cantonensis (Nematoda: Metastrongylidae) in Achatina Fulica (Mollusca: Gastropoda) from Southeast and South Brazil. *Memórias Do Inst. Oswaldo Cruz* **2010**, *105*, 938–941. [CrossRef]
7. Cowie, R.H.; Ansdell, V.; Panosian Dunavan, C.; Rollins, R.L. Neuroangiostrongyliasis: Global Spread of an Emerging Tropical Disease. *Am. J. Trop. Med. Hyg.* **2022**, *107*, 1166–1172. [CrossRef]
8. Federspiel, F.; Skovmand, S.; Skarphedinsson, S. Eosinophilic Meningitis Due to Angiostrongylus Cantonensis in Europe. *Int. J. Infect. Dis.* **2020**, *93*, 28–39. [CrossRef]
9. Wang, Q.P.; Lai, D.H.; Zhu, X.Q.; Chen, X.G.; Lun, Z.R. Human Angiostrongyliasis. *Lancet Infect. Dis.* **2008**, *8*, 621–630. [CrossRef]
10. Ansdell, V.; Wattanagoon, Y. Angiostrongylus Cantonensis in Travelers: Clinical Manifestations, Diagnosis, and Treatment. *Curr. Opin. Infect. Dis.* **2018**, *31*, 399–408. [CrossRef]
11. Solorzano, L.F.; Martini Robles, L.; Hernandez, H.; Sarracent, J.; Muzzio, J.; Rojas, L. *Angiostrongylus cantonensis*: Un parásito emergente en Ecuador. *Rev. Cuba. Med. Trop.* **2014**, *66*, 20–33.
12. Martini-Robles, L.; Gómez-Landires, E.; Muzzio-Aroca, J.; Solórzano-Álava, L. Descripcion Del Primer Foco de Transmision Natural de Angiostrongylus Cantonensis En Ecuador. In *Angiostrongylus cantonensis—Emergencia en América*; Martini-Robles, L., Dorta-Contreras, A.J., Eds.; Editorial Academia La Habana: La Habana, Cuba, 2016; pp. 209–220, ISBN 978-959-270-368-1.
13. Pincay, T.; García, L.; Narváez, E.; Decker, O.; Martini, L.; Moreira, J.M. Angiostrongiliasis Por Parastrongylus (Angiostrongylus) Cantonensis En Ecuador. Primer Informe En Sudamérica. *Trop. Med. Int Health* **2009**, *14*, S37.
14. Guerrero, M.; Vargas, F.M.; Rosero, A.R.; Nuques, M.L.; Bolaños, E.S.; Briones, M.T.; Martínez, W.Z.; Gómez, A.O. Meningitis eosinofílica por angiostrongylus cantonensis. Reporte de caso de autopsia. *Medicina* **2008**, *13*, 312–318.
15. Solórzano, L.; Sánchez-Amador, F.; Valverde, T. Angiostrongylus (Parastrongylus) cantonensis en huéspedes intermediarios y definitivos en Ecuador, 2014–2017. *Biomédica* **2019**, *39*, 370–384. [CrossRef]
16. Muzzio, J. *Hospederos Intermediarios de Angiostrongylus Cantonensis en Ecuador*; Editorial Academica Española: Saarbrücken, Germany, 2014.
17. Thiengo, S.C.; Faraco, F.A.; Salgado, N.C.; Cowie, R.H.; Fernandez, M.A. Rapid Spread of an Invasive Snail in South America: The Giant African Snail, Achatina Fulica, in Brasil. *Biol. Invasions* **2007**, *9*, 693–702. [CrossRef]
18. Moreira, V.L.C.; Giese, E.G.; Melo, F.T.V.; Simões, R.O.; Thiengo, S.C.; Maldonado-Júnior, A.; Santos, J.N. Endemic Angiostrongyliasis in the Brazilian Amazon: Natural Parasitism of Angiostrongylus Cantonensis in Rattus Rattus and R. Norvegicus, and Sympatric Giant African Land Snails, Achatina Fulica. *Acta Trop.* **2013**, *125*, 90–97. [CrossRef]
19. Qvarnstrom, Y.; Sullivan, J.J.; Bishop, H.S.; Hollingsworth, R.; da Silva, A.J. PCR-Based Detection of Angiostrongylus Cantonensis in Tissue and Mucus Secretions from Molluscan Hosts. *Appl. Environ. Microbiol.* **2007**, *73*, 1415–1419. [CrossRef] [PubMed]
20. Qvarnstrom, Y.; da Silva, A.J.C.A.; Teem, J.L.; Hollingsworth, R.; Bishop, H.S.; Graeff-Teixeira, C.; Aramburu Da Silva, A.C. Improved Molecular Detection of Angiostrongylus Cantonensis in Mollusks and Other Environmental Samples with a Species-Specific Internal Transcribed Spacer Based TaqMan Assay. *Appl. Environ. Microbiol.* **2010**, *76*, 5287–5289. [CrossRef]
21. Qvarnstrom, Y.; Xayavong, M.; da Silva, A.C.A.; Park, S.Y.; Whelen, A.C.; Calimlim, P.S.; Sciulli, R.H.; Honda, S.A.A.; Higa, K.; Kitsutani, P.; et al. Real-Time Polymerase Chain Reaction Detection of Angiostrongylus Cantonensis DNA in Cerebrospinal Fluid from Patients with Eosinophilic Meningitis. *Am. J. Trop. Med. Hyg.* **2016**, *94*, 176–181. [CrossRef]
22. Qvarnstrom, Y.; Bishop, H.S.; da Silva, A.J. Detection of Rat Lungworm in Intermediate, Definitive, and Paratenic Hosts Obtained from Environmental Sources. *Hawaii J. Med. Public Health* **2013**, *72*, 63–69.
23. Sears, W.J.; Qvarnstrom, Y.; Nutman, T.B. RPAcan3990: An Ultrasensitive Recombinase Polymerase Assay To Detect Angiostrongylus Cantonensis DNA. *J. Clin. Microbiol.* **2021**, *59*, e0118521. [CrossRef]
24. Liu, C.-Y.; Song, H.-Q.; Zhang, R.-L.; Chen, M.-X.; Xu, M.-J.; Ai, L.; Chen, X.-G.; Zhan, X.-M.; Liang, S.-H.; Yuan, Z.-G.; et al. Specific Detection of Angiostrongylus Cantonensis in the Snail Achatina Fulica Using a Loop-Mediated Isothermal Amplification (LAMP) Assay. *Mol. Cell Probes* **2011**, *25*, 164–167. [CrossRef] [PubMed]
25. Lv, S.; Zhang, Y.; Steinmann, P.; Utzinger, J.; Zhou, X.-N. The Genetic Variation of Angiostrongylus Cantonensis in the People's Republic of China. *Infect. Dis. Poverty* **2017**, *6*, 125. [CrossRef] [PubMed]

26. Liu, C.Y.; Zhang, R.L.; Chen, M.X.; Li, J.; Ai, L.; Wu, C.Y.; Zhu, X.-Q.; Lin, Q. Characterisation of Angiostrongylus Cantonensis Isolates from China by Sequences of Internal Transcribed Spacers of Nuclear Ribosomal DNA. *J. Anim. Vet. Adv.* **2011**, *10*, 593–596. [CrossRef]
27. Peng, J.; He, Z.; Zhang, S.; Lun, Z.; Wu, Z.; Fan, C.-K.; Brown, C.L.; Cheng, P.; Peng, S.; Yang, T. Phylogeography of Angiostrongylus Cantonensis (Nematoda: Angiostrongylidae) in Southern China and Some Surrounding Areas. *PLoS Negl. Trop. Dis.* **2017**, *11*, e0005776. [CrossRef]
28. Fontanilla, I.K.C.; Wade, C.M. The Small Subunit (SSU) Ribosomal (r) RNA Gene as a Genetic Marker for Identifying Infective 3rd Juvenile Stage Angiostrongylus Cantonensis. *Acta Trop.* **2008**, *105*, 181–186. [CrossRef]
29. Eamsobhana, P.; Lim, P.E.; Yong, H.S. Phylogenetics and Systematics of Angiostrongylus Lungworms and Related Taxa (Nematoda: Metastrongyloidea) Inferred from the Nuclear Small Subunit (SSU) Ribosomal DNA Sequences. *J. Helminthol.* **2015**, *89*, 317–325. [CrossRef]
30. Galtier, N.; Nabholz, B.; Glémin, S.; Hurst, G.D.D. Mitochondrial DNA as a Marker of Molecular Diversity: A Reappraisal. *Mol. Ecol.* **2009**, *18*, 4541–4550. [CrossRef]
31. Blouin, M.S. Molecular Prospecting for Cryptic Species of Nematodes: Mitochondrial DNA versus Internal Transcribed Spacer. *Int. J. Parasitol.* **2002**, *32*, 527–531. [CrossRef]
32. Yong, H.S.; Eamsobhana, P.; Song, S.-L.; Prasartvit, A. Molecular Phylogeography of Angiostrongylus Cantonensis (Nematoda: Angiostrongylidae) and Genetic Relationships with Congeners Using Cytochrome b Gene Marker. *Acta Trop.* **2015**, *148*, 66–71. [CrossRef]
33. de Almeida, L.R.; de Souza, J.G.R.; Santos, H.A.; Torres, E.J.L.; do Val Vilela, R.; Cruz, O.M.S.; Rodrigues, L.; de Jesus Pereira, C.A.; Maldonado-Júnior, A.; dos Santos Lima, W. Angiostrongylus Minasensis n. Sp.: New Species Found Parasitizing Coatis (Nasua Nasua) in an Urban Protected Area in Brazil. *Rev. Bras. De Parasitol. Veterinária* **2020**, *29*, e018119. [CrossRef]
34. Eamsobhana, P.; Lim, P.-E.; Solano, G.; Zhang, H.; Gan, X.; Yong, H.S. Molecular Differentiation of Angiostrongylus Taxa (Nematoda: Angiostrongylidae) by Cytochrome c Oxidase Subunit I (COI) Gene Sequences. *Acta Trop.* **2010**, *116*, 152–156. [CrossRef]
35. Caldeira, R.L.; Carvalho, O.S.; Mendonça, C.L.G.F.; Graeff-Teixeira, C.; Silva, M.C.; Ben, R.; Maurer, R.; Lima, W.S.; Lenzi, H.L. Molecular Differentiation of Angiostrongylus Costaricensis, A. Cantonensis, and A. Vasorum by Polymerase Chain Reaction-Restriction Fragment Length Polymorphism. *Memórias Do Inst. Oswaldo Cruz* **2003**, *98*, 1039–1043. [CrossRef]
36. Valentyne, H.; Spratt, D.M.; Aghazadeh, M.; Jones, M.K.; Šlapeta, J. The Mitochondrial Genome of Angiostrongylus Mackerrasae Is Distinct from A. Cantonensis and A. Malaysiensis. *Parasitology* **2020**, *147*, 681–688. [CrossRef] [PubMed]
37. Vitta, A.; Srisongcram, N.; Thiproaj, J.; Wongma, A.; Polsut, W.; Fukruksa, C.; Yimthin, T.; Mangkit, B.; Thanwisai, A.; Dekumyoy, P. Phylogeny of Angiostrongylus Cantonensis in Thailand Based on Cytochrome c Oxidase Subunit I Gene Sequence. *Southeast Asian J. Trop. Med. Public Health* **2016**, *47*, 377–386.
38. Monte, T.C.C.; Simões, R.O.; Oliveira, A.P.M.; Novaes, C.F.; Thiengo, S.C.; Silva, A.J.; Cordeiro-Estrela, P.; Maldonado-Júnior, A. Phylogenetic Relationship of the Brazilian Isolates of the Rat Lungworm Angiostrongylus Cantonensis (Nematoda: Metastrongylidae) Employing Mitochondrial COI Gene Sequence Data. *Parasites Vectors* **2012**, *5*, 248. [CrossRef]
39. Eamsobhana, P.; Song, S.L.; Yong, H.S.; Prasartvit, A.; Boonyong, S.; Tungtrongchitr, A. Cytochrome c Oxidase Subunit I Haplotype Diversity of Angiostrongylus Cantonensis (Nematoda: Angiostrongylidae). *Acta Trop.* **2017**, *171*, 141–145. [CrossRef] [PubMed]
40. Tokiwa, T.; Harunari, T.; Tanikawa, T.; Komatsu, N.; Koizumi, N.; Tung, K.-C.; Suzuki, J.; Kadosaka, T.; Takada, N.; Kumagai, T.; et al. Phylogenetic Relationships of Rat Lungworm, Angiostrongylus Cantonensis, Isolated from Different Geographical Regions Revealed Widespread Multiple Lineages. *Parasitol. Int.* **2012**, *61*, 431–436. [CrossRef]
41. Legislation for the Protection of Animals Used for Scientific Purposes—Environment—European Commission. Available online: https://ec.europa.eu/environment/chemicals/lab_animals/legislation_en.htm (accessed on 29 October 2020).
42. Aguiar, P.H.; Morera, P.; Pascual, J. First Record of Angiostrongylus Cantonensis in Cuba. *Am. J. Trop. Med. Hyg.* **1981**, *30*, 963–965. [CrossRef] [PubMed]
43. Ubelaker, J.E. Systematics of Species Referred to the Genus Angiostrongylus. *J. Parasitol.* **1986**, *72*, 237–244. [CrossRef]
44. Kearse, M.; Moir, R.; Wilson, A.; Stones-Havas, S.; Cheung, M.; Sturrock, S.; Buxton, S.; Cooper, A.; Markowitz, S.; Duran, C.; et al. Geneious Basic: An Integrated and Extendable Desktop Software Platform for the Organization and Analysis of Sequence Data. *Bioinformatics* **2012**, *28*, 1647–1649. [CrossRef]
45. Edgar, R.C. MUSCLE: Multiple Sequence Alignment with High Accuracy and High Throughput. *Nucleic Acids Res.* **2004**, *32*, 1792–1797. [CrossRef] [PubMed]
46. Maddison, W.P.; Maddison, D.R. Mesquite: A Modular System for Evolutionary Analysis. 2021. Available online: https://www.mesquiteproject.org (accessed on 29 October 2020).
47. Corander, J.; Marttinen, P.; Sirén, J.; Tang, J. Enhanced Bayesian Modelling in BAPS Software for Learning Genetic Structures of Populations. *BMC Bioinform.* **2008**, *9*, 539. [CrossRef] [PubMed]
48. Corander, J.; Sirén, J.; Arjas, E. Bayesian Spatial Modeling of Genetic Population Structure. *Comput. Stat.* **2008**, *23*, 111–129. [CrossRef]

49. Ronquist, F.; Teslenko, M.; van der Mark, P.; Ayres, D.L.; Darling, A.; Hohna, S.; Larget, B.; Liu, L.; Suchard, M.A.; Huelsenbeck, J.P. MrBayes 3.2: Efficient Bayesian Phylogenetic Inference and Model Choice Across a Large Model Space. *Syst. Biol.* **2012**, *61*, 539–542. [CrossRef] [PubMed]
50. Miller, M.A.; Pfeiffer, W.; Schwartz, T. Creating the CIPRES Science Gateway for Inference of Large Phylogenetic Trees. In Proceedings of the 2010 Gateway Computing Environments Workshop (GCE), New Orleans, LA, USA, 14 November 2010; pp. 1–8.
51. Rambaut, A.; Drummond, A.J.; Xie, D.; Baele, G.; Suchard, M.A. Posterior Summarization in Bayesian Phylogenetics Using Tracer 1.7. *Syst. Biol.* **2018**, *67*, 901–904. [CrossRef]
52. Leigh, J.W.; Bryant, D. POPART: Full-Feature Software for Haplotype Network Construction. *Methods Ecol. Evol.* **2015**, *6*, 1110–1116. [CrossRef]
53. Bandelt, H.-J.; Forster, P.; Röhl, A. Median-Joining Networks for Inferring Intraspecific Phylogenies. *Mol. Biol. Evol.* **1999**, *16*, 37–48. [CrossRef]
54. Rozas, J.; Ferrer-Mata, A.; Sánchez-DelBarrio, J.C.; Guirao-Rico, S.; Librado, P.; Ramos-Onsins, S.E.; Sánchez-Gracia, A. DnaSP 6: DNA Sequence Polymorphism Analysis of Large Data Sets. *Mol. Biol. Evol.* **2017**, *34*, 3299–3302. [CrossRef]
55. Tajima, F. Statistical Method for Testing the Neutral Mutation Hypothesis by DNA Polymorphism. *Genetics* **1989**, *123*, 585–595. [CrossRef]
56. Fu, Y.-X. Statistical Tests of Neutrality of Mutations Against Population Growth, Hitchhiking and Background Selection. *Genetics* **1997**, *147*, 915–925. [CrossRef]
57. Rodpai, R.; Intapan, P.M.; Thanchomnang, T.; Sanpool, O.; Sadaow, L.; Laymanivong, S.; Aung, W.P.; Phosuk, I.; Laummaunwai, P.; Maleewong, W. Angiostrongylus Cantonensis and A. Malaysiensis Broadly Overlap in Thailand, Lao PDR, Cambodia and Myanmar: A Molecular Survey of Larvae in Land Snails. *PLoS ONE* **2016**, *11*, e0161128. [CrossRef]
58. de Wit, L.A.; Ricketts, T.H. Trade and Deforestation Predict Rat Lungworm Disease, an Invasive-Driven Zoonosis, at Global and Regional Scales. *Front. Public Health* **2021**, *9*, 680986. [CrossRef]
59. Zanol, J.; Fernandez, M.A.; de Oliveira, A.P.M.; de Moraes Russo, C.A.; Thiengo, S.C. O Caramujo Exótico Invasor Achatina Fulica (Stylommatophora, Mollusca) No Estado Do Rio de Janeiro (Brasil): Situação Atual. *Biota Neotrop.* **2010**, *10*, 447–451. [CrossRef]
60. de Paiva Barçante, J.M.; Barçante, T.A.; Dias, S.R.C.; dos Santos Lima, W. Ocorrência de Achatina Fulica Bowdich, 1822 (Mollusca: Gastropoda: Achatinoidea) No Estado de Minas Gerais, Brasil. *Bol. Mus. Biol. Mello Leitão* **2005**, *18*, 65–70.
61. Teles, H.M.S.; Fontes, L.R. Implicações Da Introdução e Dispersão de Achatina Fulica Bowdich, 1822 No Brasil. *Bol. Inst. Adolfo Lutz* **2002**, *12*, 3–5.
62. Thiengo, S.C.; de Oliveira Simões, R.; Fernandez, M.A.; Maldonado-Júnior, A. Angiostrongylus Cantonensis and Rat Lungworm Disease in Brazil. *Hawai'i J. Med. Public Health* **2013**, *72*, 18–22.
63. Arruda, J.O.; Santos, L. First Record of Achatina Fulica Bowdich, 1822 (Mollusca, Achatinidae), for the State of Rio Grande Do Sul, Brazil. *Biotemas* **2022**, *35*, 1–6. [CrossRef]
64. Borrero, F.J.; Breure, A.S.H.; Christensen, C.C.; Correoso, M.; Avila, V.M. Into the Andes: Three New Introductions of Lissachatina Fulica (*Gastropoda, Achatinidae*) and Its Potential Distribution in South America. *Tentacle* **2009**, *17*, 6–8.
65. Caldeira, R.L.; Mendonça, C.L.G.F.; Goveia, C.O.; Lenzi, H.L.; Graeff-Teixeira, C.; Lima, W.S.; Mota, E.M.; Pecora, I.L.; De Medeiros, A.M.Z.; Carvalho, O.D.S. First Record of Molluscs Naturally Infected with Angiostrongylus Cantonensis (Chen, 1935) (Nematoda: Metastrongylidae) in Brazil. *Memórias Inst. Oswaldo Cruz* **2007**, *102*, 887–889. [CrossRef]
66. Lv, S.; Zhang, Y.; Liu, H.-X.; Hu, L.; Yang, K.; Steinmann, P.; Chen, Z.; Wang, L.-Y.; Utzinger, J.; Zhou, X.-N. Invasive Snails and an Emerging Infectious Disease: Results from the First National Survey on Angiostrongylus Cantonensis in China. *PLoS Negl. Trop. Dis.* **2009**, *3*, e368. [CrossRef]
67. Tirira, S.D. *Guía de Campo de Los Mamíferos Del Ecuador*; Murciélago Blanco: Quito, Ecuador, 2007; ISBN 9978-44-651-6.
68. Kosoy, M.; Khlyap, L.; Cosson, J.-F.; Morand, S. Aboriginal and Invasive Rats of Genus *Rattus* as Hosts of Infectious Agents. *Vector-Borne Zoonotic Dis.* **2015**, *15*, 3–12. [CrossRef] [PubMed]
69. Lowe, S.; Browne, M.; Boudjelas, S.; De Poorter, M. 100 of the World's Worst Invasive Alien Species: A Selection From The Global Invasive Species Database. *Aliens* **2000**, *12*, 1–12. [CrossRef]
70. Crowl, T.A.; Crist, T.O.; Parmenter, R.R.; Belovsky, G.; Lugo, A.E. The Spread of Invasive Species and Infectious Disease as Drivers of Ecosystem Change. *Front. Ecol. Environ.* **2008**, *6*, 238–246. [CrossRef]
71. Chinchio, E.; Crotta, M.; Romeo, C.; Drewe, J.A.; Guitian, J.; Ferrari, N. Invasive Alien Species and Disease Risk: An Open Challenge in Public and Animal Health. *PLoS Pathog.* **2020**, *16*, e1008922. [CrossRef]
72. Červená, B.; Modrý, D.; Fecková, B.; Hrazdilová, K.; Foronda, P.; Alonso, A.M.; Lee, R.; Walker, J.; Niebuhr, C.N.; Malik, R.; et al. Low Diversity of Angiostrongylus Cantonensis Complete Mitochondrial DNA Sequences from Australia, Hawaii, French Polynesia and the Canary Islands Revealed Using Whole Genome next-Generation Sequencing. *Parasites Vectors* **2019**, *12*, 241. [CrossRef] [PubMed]
73. Monte, T.C.C.; Gentile, R.; Garcia, J.S.; Mota, E.; dos Santos, J.N.; Maldonado-Júnior, A. Brazilian Angiostrongylus Cantonensis Haplotypes, Ac8 and Ac9, Have Two Different Biological and Morphological Profiles. *Mem. Do Inst. Oswaldo Cruz* **2014**, *109*, 1057–1063. [CrossRef] [PubMed]

74. Tokiwa, T.; Hashimoto, T.; Yabe, T.; Komatsu, N.; Akao, N.; Ohta, N. First Report of Angiostrongylus Cantonensis (Nematoda: Angiostrongylidae) Infections in Invasive Rodents from Five Islands of the Ogasawara Archipelago, Japan. *PLoS ONE* **2013**, *8*, e70729. [CrossRef]
75. Rael, R.C.; Peterson, A.C.; Ghersi-Chavez, B.; Riegel, C.; Lesen, A.E.; Blum, M.J. Rat Lungworm Infection in Rodents across Post-Katrina New Orleans, Louisiana, USA. *Emerg. Infect. Dis.* **2018**, *24*, 2176–2183. [CrossRef]

pathogens

MDPI

Article

The Global Spread Pattern of Rat Lungworm Based on Mitochondrial Genetics

Xia Tian [1,2,3,4,5], Shen Chen [1,2,3,4,5], Lei Duan [1,2,3,4,5], Yingjun Qian [1,2,3,4,5], Hongmei Li [1,2,3,4,5] and Shan Lv [1,2,3,4,5,*]

1 National Institute of Parasitic Diseases, Chinese Center for Disease Control and Prevention, Shanghai 200025, China; tx09871@163.com (X.T.); nematode@163.com (S.C.); duanlei@nipd.chinacdc.cn (L.D.); qianyj@nipd.chinacdc.cn (Y.Q.); lihm@nipd.chinacdc.cn (H.L.)
2 Key Laboratory on Parasite and Vector Biology, National Health Commission, Shanghai 200025, China
3 WHO Collaborating Centre for Tropical Diseases, Shanghai 200025, China
4 National Center for International Research on Tropical Diseases, Ministry of Science and Technology, Shanghai 200025, China
5 Chinese Center for Tropical Diseases Research, Shanghai 200025, China
* Correspondence: lvshan@nipd.chinacdc.cn

Abstract: Eosinophilic meningitis due to rat lungworm, *Angiostrongylus cantonensis*, is a global public health concern. Human cases and outbreaks have occurred in the new endemic areas, including South America and Spain. The growing genetic data of *A. cantonensis* provides a unique opportunity to explore the global spread pattern of the parasite. Eight more mitochondrial (mt) genomes were sequenced by the present study. The phylogeny of *A. cantonensis* by Bayesian inference showed six clades (I–VI) determined by network analysis. A total of 554 mt genomes or fragments, which represented 1472 specimens of rat lungworms globally, were used in the present study. We characterized the gene types by mapping a variety of mt gene fragments to the known complete mt genomes. Six more clades (I2, II2, III2, V2, VII and VIII) were determined by network analysis in the phylogenies of *cox*1 and *cyt*b genes. The global distribution of gene types was visualized. It was found that the haplotype diversity of *A. cantonensis* in Southeast and East Asia was significantly higher than that in other regions. The majority (78/81) of samples beyond Southeast and East Asia belongs to Clade II. The new world showed a higher diversity of Clade II in contrast with the Pacific. We speculate that rat lungworm was introduced from Southeast Asia rather than the Pacific. Therefore, systematic research should be conducted on rat lungworm at a global level in order to reveal the scenarios of spread.

Keywords: *Angiostrongylus cantonensis*; *Angiostrongylus mackerrasae*; *Angiostrongylus malaysiensis*; rat lungworm; mitochondrial gene; distribution; phylogeny

Citation: Tian, X.; Chen, S.; Duan, L.; Qian, Y.; Li, H.; Lv, S. The Global Spread Pattern of Rat Lungworm Based on Mitochondrial Genetics. *Pathogens* **2023**, *12*, 788. https://doi.org/10.3390/pathogens12060788

Academic Editor: José María Alunda

Received: 23 April 2023
Revised: 15 May 2023
Accepted: 29 May 2023
Published: 31 May 2023

1. Introduction

Angiostrongylus cantonensis is the major cause of human neural angiostrongyliasis [1]. Infection with this nematode often results in eosinophilic meningitis (EM) and other disorders of the central nervous system [2]. Humans are vulnerable to infection because of the varied contact opportunities with mollusk intermediate hosts and paratenic hosts. EM occurs following ingestion of undercooked or raw snails, slugs, freshwater shrimp or crabs, lizards, and even contaminated vegetable salad with mollusk slime [3].

A. cantonensis was first discovered from the pulmonary arteries and heart of rats in southern China in 1933 [4]. Human EM due to *A. cantonensis* was first reported in 1945 [5], followed by EM outbreaks in Pohnpei, of the Caroline Islands of the Pacific between 1947 and 1948 [6]. However, the health concern was not noted in the early 1960s when an outbreak occurred in Tahiti [7]. The subsequent surveys indicated that EM due to

A. cantoensis was endemic in Southeast Asia and the Pacific [8]. Cases reported in recent years indicate that *A. cantoensis* constitutes an emerging zoonosis worldwide [9–11].

The expansion of the endemic range of *A. cantonensis* has been, to an extent, attributed to biological invasion. The common definitive host, the black rat *Rattus rattus* and the Norway rat *R. norvegicus*, are listed as the world's top invasive species [12]. The growing shipping industry, particularly for cargo, drove the global spread of the two Rattus species and thus *A. cantonensis*. The intermediate host snail, *Achatina fulica*, was thought to play a key role in the spread of *A. cantonensis* in World War II, particularly in the Pacific islands [13]. Indeed, the land snail occupied the islands quickly in the 1940s and 1950s [14]; however, some human cases attributed to the consumption of *A. fulica* were reported earlier [15]. The snail has also recently begun to be considered an important factor in the occurrence of *A. cantonensis* in South America [16]. Another global invasive snail species is *Pomacea canaliculata*, which has been widely distributed in Southeast and East Asia since the 1980s [17]. The freshwater snail has driven the emergence of human neural angiostrongyliasis in the region [18]. The continuous invasion and re-introduction of definitive and intermediate hosts make the global spread of *A. cantonensis* more complex.

Rat lungworm commonly refers to *A. cantonensis*. However, rat lungworm also includes *A. mackerrasae* and *A. malaysiensis*, which were not distinguished from *A. cantonensis* until the late 1960s [19,20]. Although they have a nearly consistent life cycle, their geographical range and preference for a definitive host are different. In contrast with the global distribution of *A. cantonensis*, *A. mackerrasae* is confined to the east coast of Australia [21] and *A. malaysiensis* is mainly distributed in Thailand, Malaysia and Indonesia [19,22]. The most common definitive host of *A. mackerrasae* is the bush rat *R. fuscipes* and swamp rat *Rattus lutreolus* [21], while *A. malaysiensis* prefers the Malaysian field rat *R. tiomanicus* and the ricefield rat *R. argentiventer* [23]. Nevertheless, the two species have also been discovered from the global invasive *R. norvegicus*. Therefore, it is possible that the two species of *Angiostrongylus* may be distributed beyond the original regions. In addition, the diagnosis of most infections in human and animal was based on larval morphology and immunology, the infections due to *A. mackerrasae* and *A. malaysiensis* were probably misdiagnosed. For example, a recent report has shown through molecular evidence that *A. malaysiensis* can infect humans [24].

The genetic data of rat lungworms are increasingly prevalent around the world, providing an opportunity to reveal the global spread pattern of *A. cantoensis* and check the possibility of a range expansion for *A. mackerrasae* and *A. malaysiensis*. However, the molecular markers are diverse, with a focus on mitochondrial (mt) genes, including cytochrome c oxidase 1 gene (*cox*1) [25,26], cytochrome b gene (*cyt*b) [27], NADH dehydrogenase 1 gene (*nad*1) [28], ribosomal subunit RNA gene [29], and even complete mt genomes [30]. In the present study, we will characterize the available complete mt genomes and map the known gene fragments to the mt genomes so that we can categorize the samples from multiple published works. Finally, the global distribution patterns of the gene types will be revealed and the global spread pattern of *A. cantoensis* will be inferred.

2. Materials and Methods

2.1. Mitochondrial Genomes

The present study will provide eight full mt genomes of *A. cantonensis*. The adult worms of *A. cantonensis* were collected from the pulmonary arteries of wild rats during the first national survey in China [31], or from Sprague–Dawley rats that were infected by third stage larvae from snails or slugs. Nine female worms from seven collecting sites were used for sequencing mt genomes (Table 1). One of the full sequences had been published (GQ398121). The other eight were characterized by the same methods and used for the present study [32]. In addition, the mt genomes of *A. cantonensis*, *A. malaysiensis* and *A. mackerrasae* that are deposited in GenBank were also included in the present analysis.

Table 1. The mitochondrial genomes of *Angiostrongylus* used in phylogenetic analysis.

Species	Access Number	Location	Altitude	Longitude	References
A. cantonensis	AP017672	Taipei, China	25.0329	121.5655	[30]
	CL_Acan	Changle, China	25.9313	119.6288	this study
	DF1_Acan	Dongfang, China	19.0535	108.6521	this study
	DF2_Acan	Dongfang, China	19.0535	108.6521	this study
	GQ398121	Lianjiang, China	26.2052	119.5212	[32]
	KT947978	Thailand			[26]
	MK570629	Tenerife, Spain	28.2916	−16.6291	[30]
	MK570630	Hawaii, USA	19.6310	−156.0072	[30]
	MK570631	Mosman, Australia	−33.8293	151.2442	[30]
	MK570632	Fatu Hiva, French Polynesia	−17.6427	−149.4347	[30]
	NA_Acan	Nanao, China	23.4533	117.0971	this study
	RL_Acan	Ruili, China	23.9477	97.7854	this study
	SC_Acan	Shangchuan, China	21.6613	112.8016	this study
	SY_Acan	Sanya, China	18.3333	109.4333	this study
	ZX_Acan	Zixing, China	26.0365	113.2463	this study
A. mackerrasae	MN793157	Brisbane, Australia	−27.4620	153.0203	[33]
A. malaysiensis	KT186242	Thailand			[34]
	KT947979	Kuala Lumpur, Malaysia	3.1201	101.6545	[26]
A. vasorum	JX268542	Australia			[35]
A. costaricensis	GQ398122	Brazil			[32]

2.2. Mitochondrial Gene Sequences

Since our analysis strategy is to map the mt genes to the known complete mt genomes and then determine the types of each sequence, all mt gene sequences of *A. cantonensis*, *A. malaysiensis* and *A. mackerrasae* were collected from GenBank and used for the present study. Meanwhile, the collection information of specimens, e.g., location, year, species, gene name, sample size were also collected. For the sequences which had been published in papers but not yet submitted to GenBank, we collected the sequences directly from the papers including Supplementary Materials.

2.3. Phylogenetic Analysis

The complete mt genomes of *A. cantonensis*, *A. malaysiensis* and *A. mackerrasae* were used to construct phylogeny with the *A. vasorum* (JX268542) and *A. costaricensis* (GQ398122) as outgroups. A phylogenetic tree was constructed under Bayesian inference, performed in MrBayes version 3.2.7 [31]. Prior to Bayesian inference, the best fit nucleotide substitution model (GTR+G) for the dataset was determined using a hierarchical likelihood ratio test in jModeltest version 2 [32]. The posterior probabilities were calculated using Markov chain Monte Carlo (MCMC) simulations. At the end of this run, the average standard deviation of split frequencies was below 0.01, and the potential scale reduction factor was reasonably close to 1.0 for all parameters. A consensus tree was visualized and edited by Mesquite version 3.70 [36] or FigTree version 1.4.4 (http://tree.bio.ed.ac.uk/software/Figtree/; accessed on 28 May 2023).

The group of complete mt genomes was also submitted to network analysis by TCS version 1.21 [36]. Subgroups were determined with a connection limit of 1% or 132 steps and considered as the clades in the phylogenetic tree mentioned above.

The mt gene sequences downloaded from GenBank and papers were first categorized by the gene name. The sequences of the same gene were mapped to the complete mt genomes by alignment in clustal X. The sequences might be divided into subgroups when the overlapped length was less than 50%. After being trimmed as necessary, the grouped sequences were submitted to DnaSP version 5 [37] and the haplotypes were determined. The phylogenetic trees based on the haplotypes were inferred following the guide described above.

Some studies characterized the sequence of different mt genes of the same specimen, e.g., *cox*1 and *cyt*b. Only one gene for a single specimen was taken into analysis. The gene *cox*1 was of priority because it was used broadly, followed by *cyt*b and small ribosomal RNA gene. Large ribosomal RNA gene and nad1 were excluded because *cox*1 and small ribosomal RNA gene from the same specimens were included.

The gene fragments deduced from the 18 whole mt sequences of *A. cantonensis*, *A. malaysiensis* and *A. malaysiensis* were included as index markers in the individual gene phylogenetic trees. Each haplotype was assigned to a specific clade as delineated by whole mt genomes according to the index markers. New clades could be defined if the haplotypes in the clades were separated by network analysis with a connection limit of 1%, the same threshold as in analysis of full mt genomes. If a new clade was clustered with a specific known clade inferred according to the full mt genomes, the new clade would have the same name with a suffix of 2. Otherwise, a new name would be given to the new clade.

2.4. Mapping the Distribution Pattern

The coordinates of specimen collecting sites were either directly collected from the published papers or determined by Google Earth by inputting the location names which were extracted from the papers and GenBank. Each gene sequence from known locations was assigned to a specific clade determined in phylogenetic analysis.

The geographic distributions of gene types were mapped and visualized in a geographical information system (ArcGIS version 10.1). We merged some collecting sites for better display when the distance among them was less than 0.8 degrees at the global level and less than 0.5 degrees at the subregion level.

3. Results

A total of 18 mt genomes of rat lungworms were used in the analysis, including 10 from GenBank and 8 from the present study (Table 1, File S1). The genetic distance among three species of rat lungworm was around 0.12, while the genetic distance among various strains of *A. cantonensis* ranged from 0.001 to 0.056 (Table 2). The phylogenetic and network analyses showed that 15 mt genomes of *A. cantonensis* collapsed into six clades, encoded as clade I to clade VI (Figure 1). One genome (KT186242), previously identified as *A. cantonensis*, had closer genetic relation to *A. malaysiensis* (KT947979) according to the present study. Another genome (MN793157) belonged to *A. mackerrasae*.

The individual gene phylogenetic trees of *cox*1, *nad*1 and small ribosomal RNA genes from 18 complete mt genomes were re-constructed by Bayesian inference (File S2). When incorporated into the individual gene phylogenetic trees, the gene fragments deduced from the whole mt sequences all remained in their original clades, as expected.

A total of 554 mt genomes or fragments, which represented 1472 specimens of rat lungworms globally (Table 3), were used in the present study. The specimens were collected from 224 sites. According to our analysis one mt genome (KT186242), 13 *cox*1 sequences and 64 *cyt*b sequences were *A. malaysiensis* and had previously been identified as *A. cantonensis*.

There are 327 *cox*1 sequences, including 160 fragments from GenBank, 18 mt genomes and 149 sequences from our previous study. One sequence (LC515249) was excluded due to the lack of overlap with any other sequence. The rest of the 326 sequences were divided into two subgroups according to the criteria of an overlap of less than 50%. The first subgroups included 308 sequences, which belonged to 112 haplotypes. All six of the clades (I ~ VI) based on the complete mt genomes were observed in the phylogenetic tree based on the first subgroups of *cox*1 genes. Meanwhile, another four clades were identified and named as Clade I-2, Clade II-2, Clade III-2 and Clade V-2 based on their relations to the six known clades (Figure 2). The second subgroup included 18 sequences, belonging to 11 haplotypes. All the sequences fell into Clade II (Figure 3).

Table 2. Genetic distance based on complete mitochondrial genomes.

	MN793157	KT186242	KT947979	CL_Acan	NA_Acan	GQ398121	SY_Acan	ZX_Acan	MK570632	MK570630	MK570629	AP017672	MK570631	KT947978	RL_Acan	DF1_Acan	DF2_Acan
KT186242	0.133																
KT947979	0.133	0.002															
CL_Acan	0.116	0.124	0.124														
NA_Acan	0.116	0.125	0.124	0.001													
GQ398121	0.117	0.125	0.125	0.002	0.002												
SY_Acan	0.115	0.123	0.123	0.008	0.007	0.009											
ZX_Acan	0.116	0.123	0.123	0.015	0.015	0.016	0.014										
MK570632	0.118	0.125	0.125	0.035	0.035	0.036	0.034	0.034									
MK570630	0.118	0.124	0.124	0.034	0.034	0.035	0.033	0.033	0.001								
MK570629	0.118	0.123	0.123	0.034	0.034	0.035	0.033	0.034	0.011	0.010							
AP017672	0.118	0.123	0.123	0.034	0.034	0.035	0.033	0.034	0.011	0.010	0.001						
MK570631	0.118	0.123	0.123	0.034	0.034	0.035	0.033	0.033	0.011	0.010	0.001	0.001					
KT947978	0.118	0.124	0.124	0.034	0.034	0.035	0.033	0.033	0.018	0.017	0.016	0.016	0.016				
RL_Acan	0.119	0.125	0.125	0.034	0.034	0.035	0.033	0.033	0.019	0.019	0.017	0.017	0.017	0.009			
DF1_Acan	0.118	0.124	0.124	0.035	0.035	0.036	0.033	0.034	0.030	0.029	0.029	0.029	0.029	0.029	0.030		
DF2_Acan	0.117	0.124	0.124	0.035	0.035	0.036	0.033	0.034	0.030	0.029	0.029	0.029	0.029	0.028	0.030	0.001	
SC_Acan	0.117	0.126	0.126	0.051	0.052	0.052	0.051	0.053	0.056	0.056	0.055	0.056	0.055	0.055	0.055	0.056	0.056

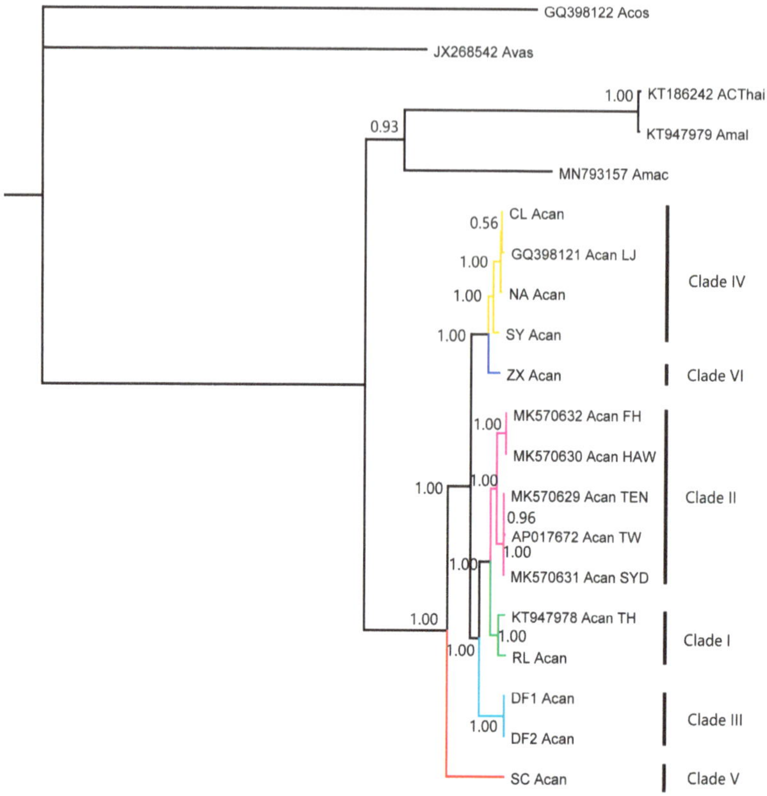

Figure 1. The phylogeny of rat lungworm inferred from mitochondrial genomes, with *A. vasorum* (JX268542) and *A. costaricensis* (GQ398122) as outgroups.

Table 3. Mitochondrial gene sequences used in the present study.

Types	*A. cantonensis*	*A. mackerrasae*	*A. malaysiensis*	Total
mt genome	8 (16 *)	1 (1)	1 (1)	10 (18)
*cox*1	144 (426 #)	5 (5)	11 (23)	160 (454)
*cyt*b	177 (655 $)	0 (0)	76 (76)	253 (731)
*nad*1	0 (130)	0 (0)	0 (0)	0 (130)
SSU	13 (13)	0 (0)	53 (53)	66 (66)
LSU	12 (12)	0 (0)	53 (53)	65 (65)
Total	354 (1252)	6 (6)	194 (206)	554 (1472)

Note: the figures inside and outside brackets represent the number of sequences from GenBank and the number of samples, respectively. SSU: small ribosomal RNAS gene; LSU: large ribosomal RNAS gene. * including 8 genomes from the present study and one genome (KT186242) of *A. malaysiensis* mistaken for *A. cantonensis*. # including 13 sequences of *A. malaysiensis* mistaken for *A. cantonensis* $ including 64 sequences of *A. malaysiensis* mistaken for *A. cantonensis*.

Ten *cyt*b sequences from the same specimens of *cox*1 were excluded in the analysis. The rest of the 195 *cyt*b sequences, including 177 from GenBank and 18 from complete mt genomes, collapsed into 67 haplotypes. Eight clades were found in the phylogenetic analysis (Figure 4), including six that were determined according to complete mt genomes and two distinctive clades named Clade VII and VIII. The average genetic distance of sequences in Clade VIII to the other clades of *A. cantonensis* was over 5% (Table S1). The difference could be as high as 7%.

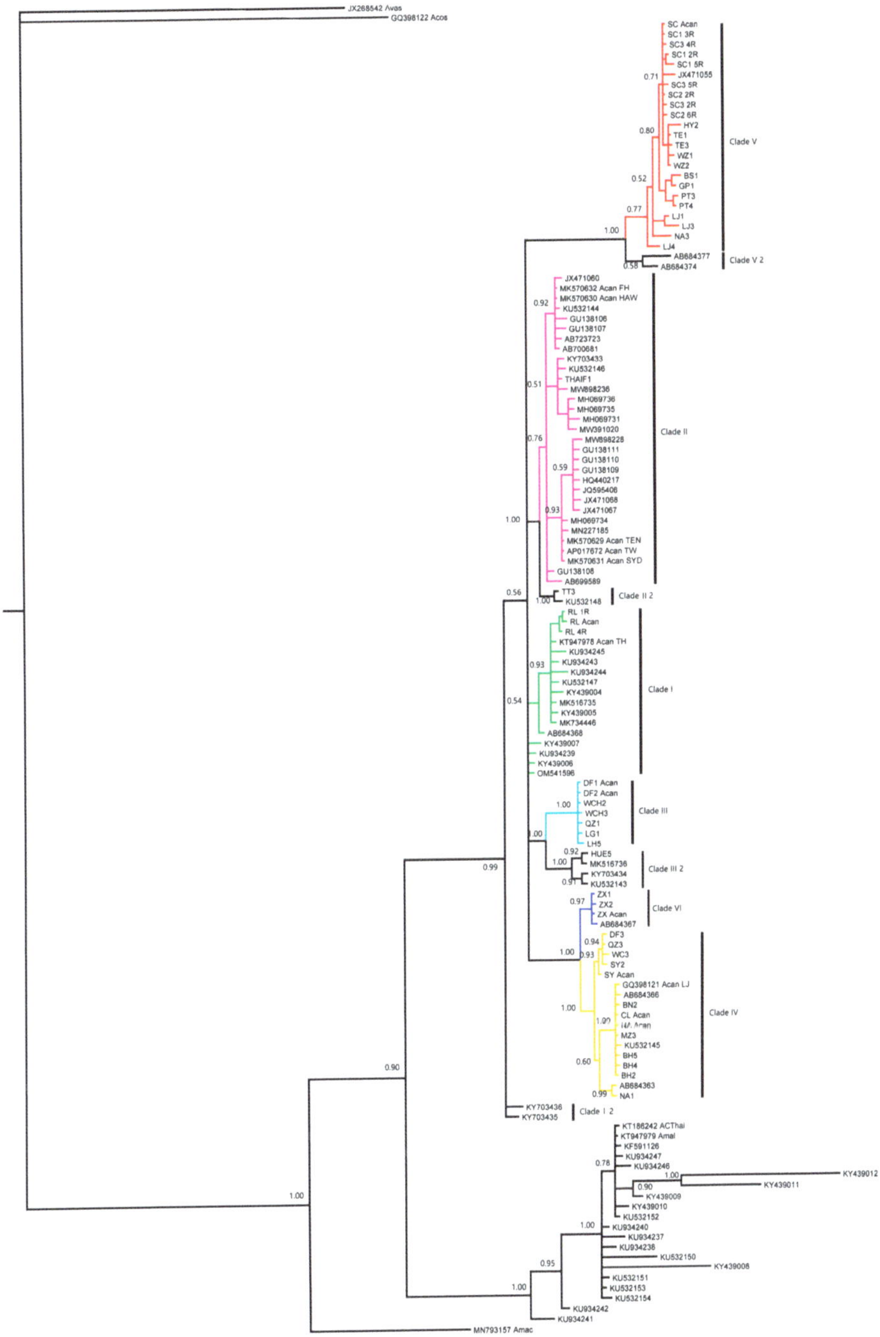

Figure 2. The phylogeny of rat lungworm inferred via the first group of *cox*1 haplotypes. The *cox*1 fragments of eighteen available full mitochondrial genomes are included as index of clades. The code and branch color is consistent with Figure 1. Four new clades, i.e., Clade I-2, Clade II-2, Clade III-2, and Clade V-2, are identified based on the network analysis.

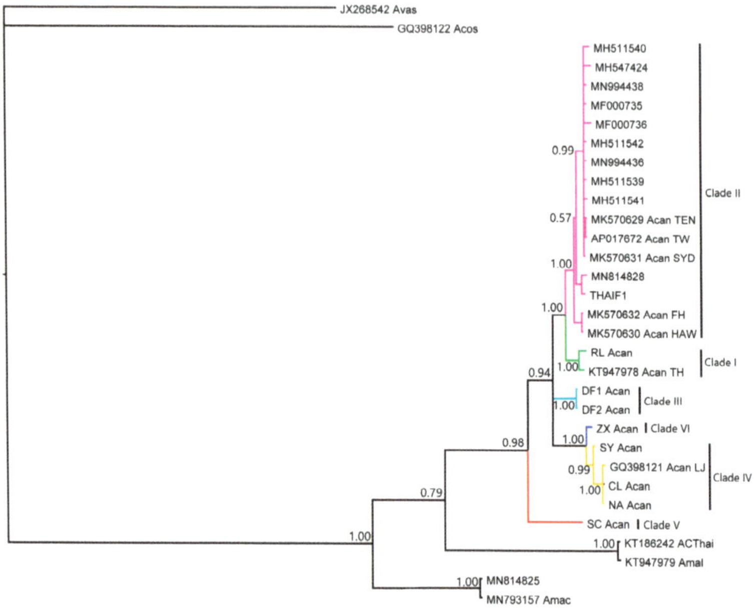

Figure 3. The phylogeny of rat lungworm inferred via the second group of *cox*1 haplotypes. The gene sequences used in this phylogeny showed <50% overlap with the sequences used in Figure 2.

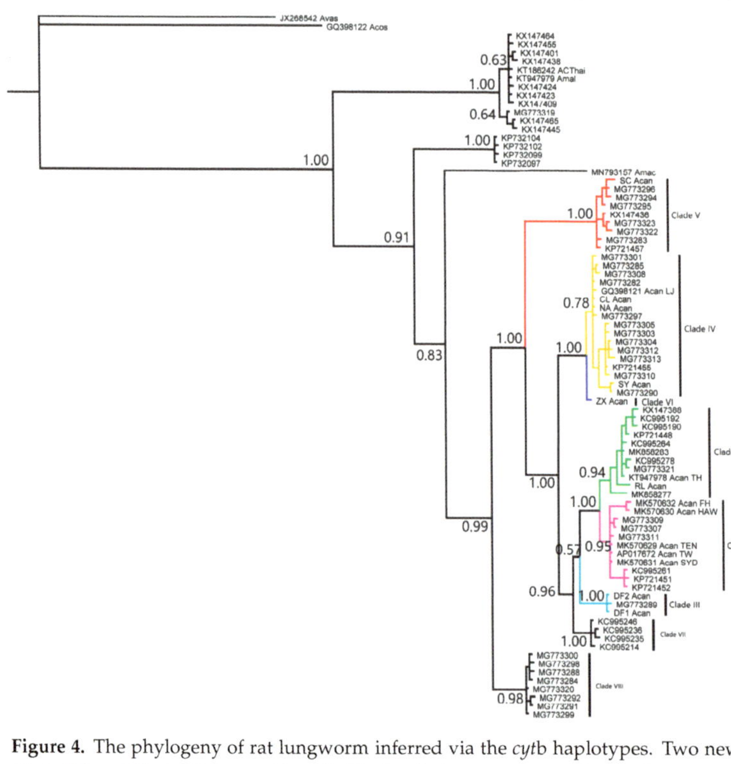

Figure 4. The phylogeny of rat lungworm inferred via the *cytb* haplotypes. Two new clades, i.e., Clade VII and Clade VIII, were identified in the phylogeny compared with Figure 1.

Twelve sequences of small and large ribosomal RNA genes were from the same specimens. Taking into consideration the smaller ribosomal subunit gene (ON747257), 13 small ribosomal RNA genes were used in the analysis. The sequences formed five haplotypes. According to network analysis, three haplotypes belonged to Clade II (Figure 5), while another two haplotypes fell into Clade I and III, respectively.

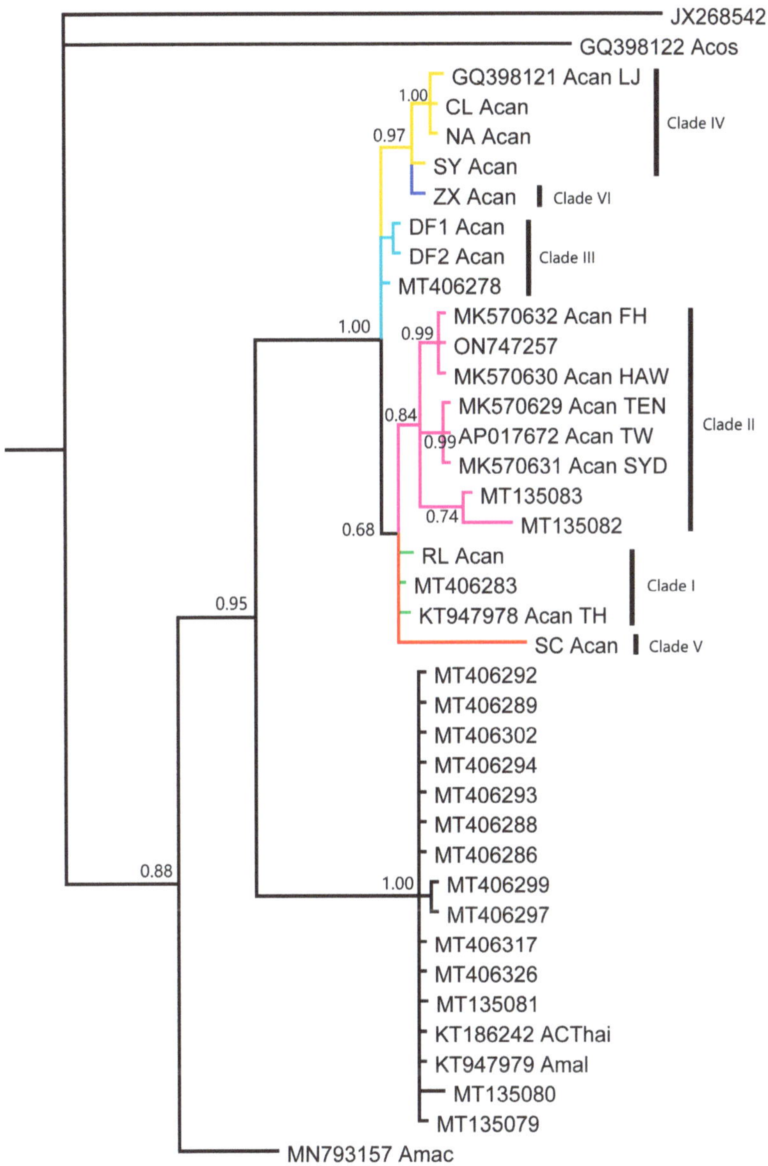

Figure 5. The phylogeny of rat lungworm inferred via the mitochondrial small ribosomal RNA gene.

The gene types (clades) based on phylogenetic and network analyses were mapped in the world (Figure 6, Table S2). A much higher diversity of clades was observed in Southeast and East Asia. In contrast, almost all 81 specimens of *A. cantonensis* beyond the region belonged to Clade II, except for two (Clade IV) from Hawaii and one (Clade

V) from Rio de Janeiro. In Asia, the Annamite range might be considered the genetic barrier for *A. cantonensis* (Figure 7). The gene types of Clade III, IV, V, V-2, VI and VIII were mainly distributed east of the Annamite range. On the contrary, Clade I, III-2 and VII were commonly found to the west of the Annamite range. *A. malaysiensis* was commonly found in in the west of Annamite range. However, it was also discovered in Taiwan according to the present analysis. *A. mackerrasae* was only observed in Australia.

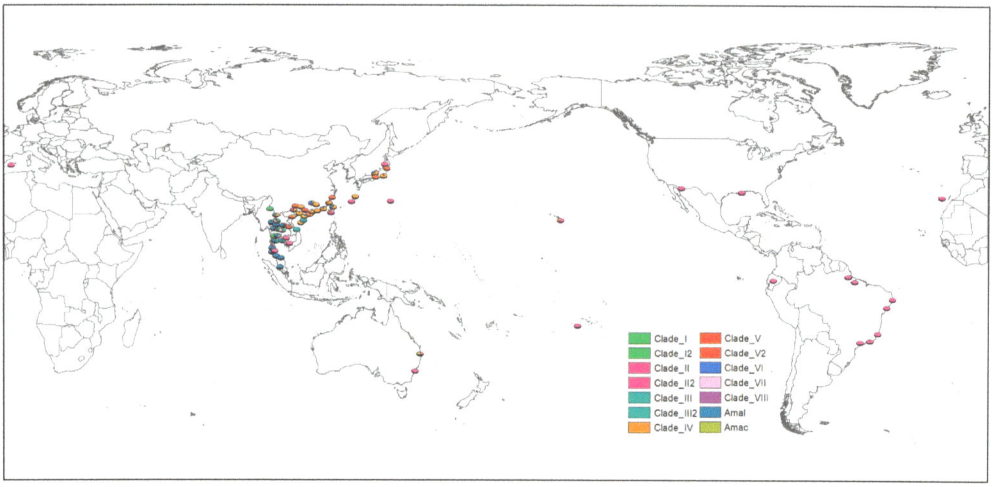

Figure 6. The global distribution of gene types of rat lungworm. The sites with a distance less than 0.8 degrees were combined and displayed at the geometric center. Amal: *A. malaysiensis*, Amac: *A. mackerrasae*.

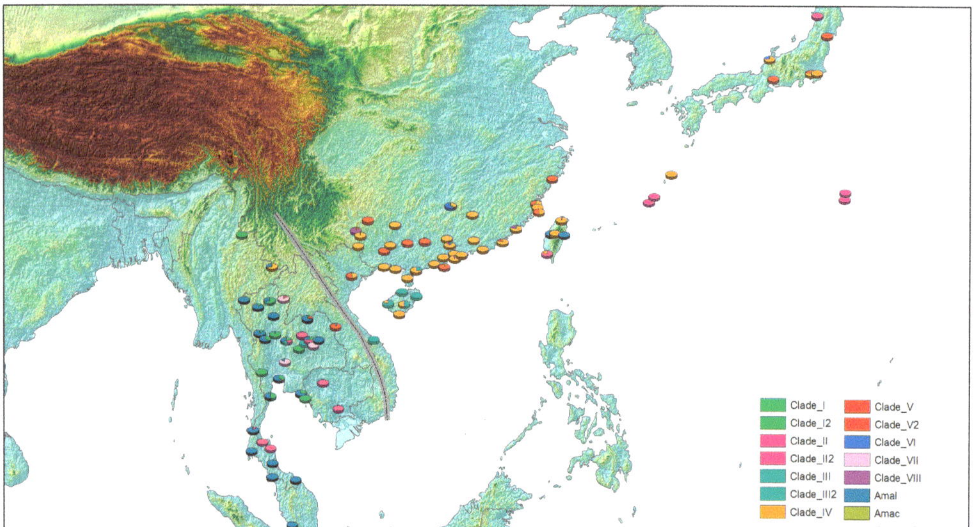

Figure 7. The distribution of gene types of rat lungworm in Southeast and East Asia. The sites with a distance of less than 0.5 degrees were combined and displayed at the geometric center. Amal: *A. malaysiensis*, Amac: *A. mackerrasae*.

The *cox*1 sequence types in Clade II were further categorized into ten subgroups (Figure 8). According to the geographic distribution of the subgroups, the haplotype diversity beyond Southeast and East Asia was 0.944, which was higher than that of 0.4641 in Asia (Figure 9). Clade II was not commonly observed in Asia according to available data. For example, Clade II was not discovered in the mainland of China and the northern part of the Greater Mekong subregion, though intensive sampling was undertaken and the high genetic diversity of *A. cantonensis* was observed in the region. Another feature of clade distribution in Asia was that none of the seven clades, except for Clade II-G, was simultaneously observed in two or more sites.

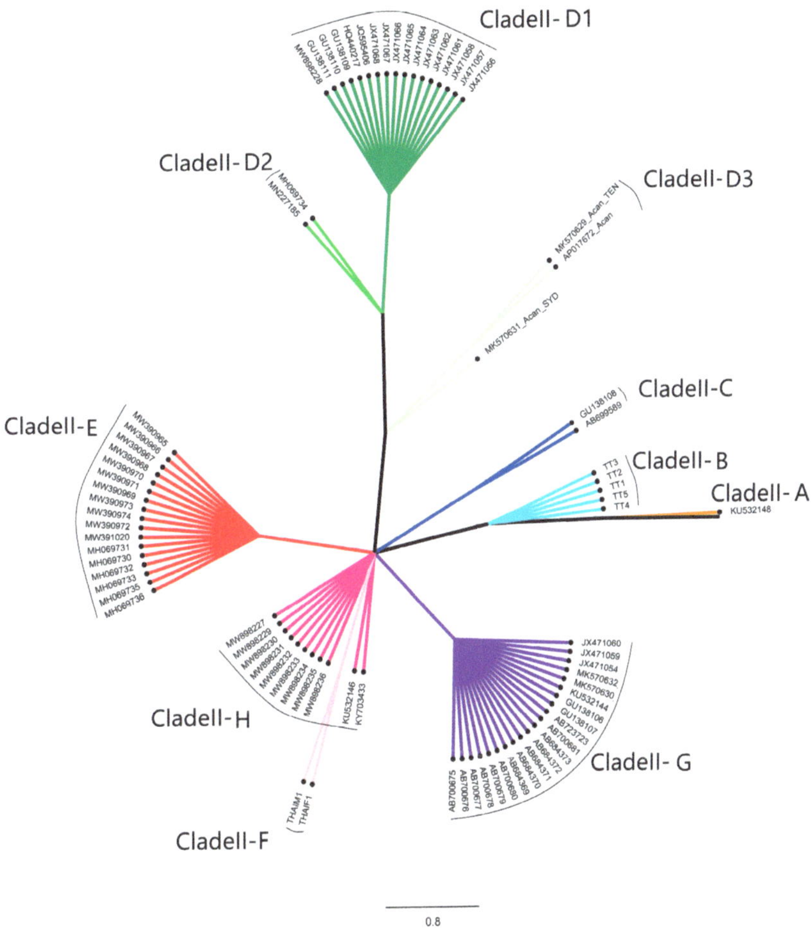

Figure 8. The phylogenetic relation of Clade II (*cox*1) subgroups.

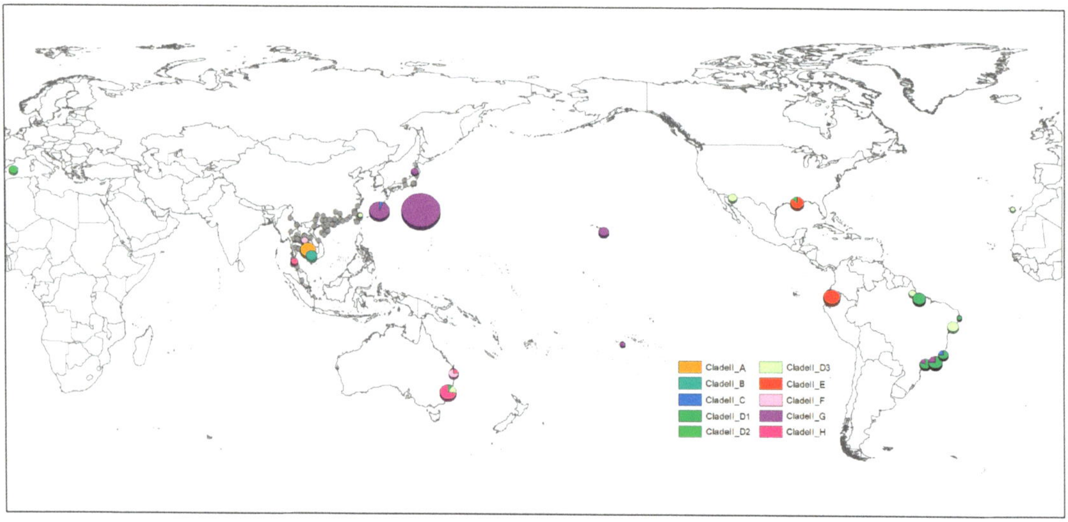

Figure 9. The global distribution of Clade II (*cox*1) with sample size. Grey dot indicate the distribution of other clades.

Clade II-G is the most common type, accounting for 48.57% of Clade II and found in seven sites in the world, five of which were located in the Pacific islands. Meanwhile the clade was also extended to the southeast shore area of Brazil. In contrast with the Pacific, the new world shows a higher diversity of clades, including the predominant clades II-D and II-E. Although the clade II-D1 and clade II-D3 were observed on Taiwan island and eastern Australia, 90% of the samples of Clade II-D were from the new world and Spain, while Australia shared two clades with Southeast Asia and South America.

4. Discussion

A total of 18 complete mt genomes of rat lungworms are currently available. Only one mt genome is *A. mackerrasae* and *A. malaysiensis*, respectively. The genetic distance among the species is over 0.11, while that among the different geographical strains within species is less than 0.06. Therefore, the species of rat lungworm could be definitively distinguished based on genetic distance. The genetic distance of the mt genome (KT186242), previously identified as *A. cantonensis*, was over 0.11 to any known strain of *A. cantonensis*. In contrast, the genome is almost identical to the mt genome (KT947979) diagnosed as *A. malaysiensis*. Therefore, KT186242 should be *A. malaysiensis*.

There were two hypotheses about the origin of *A. cantonensis*. Earlier opinion indicated that the parasite originated from Africa, which was supported by the fact that the discovery of *A. cantonensis* coincided with the spread of the African land snail *A. fulica* to Southeast Asia [38]. However, the theory of Asian origin later became popular. It was thought that the parasite was originally endemic to Southeast Asia and spread by the shipping rats, *R. rattus* and *R. norvegicus*, due to extensive traveling [13]. Our results show a much higher genetic diversity of *A. cantonensis* in Southeast and East Asia, supporting the latter theory.

Although intensive sampling occurred in China, Japan and the Great Mekong subregion, the common gene types (e.g., Clade II-D and Clade II-E shown in Figure 9) in America were rarely found in the region mentioned above. It was therefore believed that the gene types must be from anywhere else in Southeast Asia. Of note, Clade II was genetically close to Clade I. The latter was found only in Thailand and Myanmar, where *A. malaysiensis* co-occurred. The phenomenon of significant correlation between genetic and geographical distances is commonly observed in population genetics [39]. Therefore, Clade II was possibly from either Myanmar or the lower reach of the Mekong River. Similarly, Clade V showed longer genetic distance from any other clade within *A. cantonensis*, though the clade co-occurred with clade IV in China and Japan. We did not obtain any genetic evidence from the Philippines, but the country holds the potential to harbor the Clade V as a local gene type. A total of 124 *cox*1 samples from 24 collecting sites were available in Japan, and they consisted of 4 clades. However, the haplotype diversity of the samples was very low compared with Southeast Asia and the south of China [28,40]. Only seven haplotypes were identified, including three unique haplotypes within a single sample. Therefore, *A. cantonensis* in Japan might have been introduced from Southeast Asia and/or the south of China.

Our results show that there might be a genetic barrier between the Greater Mekong subregion and the south of China. The geographical isolation due to the Annamite range located in Vietnam may play a key role in the genetic divergence of *A. cantonensis*. Previous studies have indicated different evolutionary trajectories on western and eastern sides of the Annamite Mountain range [41]. The case of liver fluke also supports this opinion. Two species of liver fluke, i.e., *Clonorchis sinensis* and *Opisthorchis viverrini*, are widely distributed in the region. The former is distributed northeast of the Annamite range, while the latter is endemic in the southwest [42,43].

Clade II is the overwhelming gene type beyond Southeast and East Asia except for a small number of samples of Clade IV and Clade V in Hawaii and Rio de Janeiro, respectively. The conclusive global spread route of rat lungworm could not be established based on the present available genetic data. However, our findings imply that the gene types on the Pacific islands, where EM outbreaks occurred between the 1940s and 1960s, show identical and low diversity, which implies that its re-introduction after the Pacific War is less plausible [13]. Australia is the only country where *A. cantonensis* was reported earlier [21]. Our findings show there was no relation between Australia and the Pacific based on the available evidence, while the majority of samples are genetically close to those from Thailand. The new world, recently identified as endemic areas, probably showed a distinct haplotype structure from the Pacific and hence might have been directly introduced to the rat lungworm from Southeast Asia.

Compared with the global distribution of *A. cantonensis*, *A. mackerrasae* and *A. malaysiensis* are endemic locally. According to our results *A. mackerrasae* was only found on the eastern shore of Australia. Although *A. mackerrasae* was also discovered from *R. norvegicus*, it shows a higher susceptibility to local Australian rodents, e.g., *R. fuscipes*, *R. lutreolus* and *Melomys cervinipes* [21]. Therefore, *A. mackerrasae* was not endemic beyond the original region. *A. malaysiensis* showed a wider range of definitive hosts than *A. mackerrasae*. Although *A. malaysiensis* seems more susceptible to *R. tiomanicus* and *R. argentiventer*, the parasite shared 80% of definitive hosts with *A. cantonensis*, including the most invasive *R. norvegicus* and *R. rattus* (unpublished data). Our present study indicates that *A. malaysiensis* had gone beyond the original region and established local transmission in eastern Taiwan, since the worms had been isolated from the intermediate host *A. fulica* [44,45].

In order to reveal the complex spread pattern of *A. cantonensis*, we suggest the following research priorities. First, more mt genomes of rat lungworm should be characterized. Our present study identified new clades with complete mt genome sequences not yet available. Second, we need to fill the gap of genetic information in the Philippines and Indonesia, and even Myanmar. Since these countries are neighboring to the Greater Mekong subregion and East Asia, the mt genetic information of rat lungworms will be invaluable in constructing the phylogeny. Unfortunately, the large-scale survey has not been conducted and the mt genetic information has been lacking to date, though rat lungworms have been reported in the Philippines and Indonesia since the 1950s. Third, systemic sampling around the world is proposed to update either rare specimens or a small number of collecting sites beyond the Greater Mekong subregion and East Asia. In addition, the types of host animals, i.e., mollusks and rodents, were different in previous studies. It should be noted that the host preferences of species or strains of rat lungworm may cause a bias, and hence lead to false conclusions.

5. Conclusions

We showed an integrated global distribution of gene types by mapping various fragments to the known complete mt genomes. The new endemic areas including the new world and Spain showed different compositions of gene types to Southeast and East Asia and even the Pacific. Therefore, we need to conduct systematic research on rat lungworm at a global level in order to reveal the scenarios of spread.

Supplementary Materials: The following supporting information can be downloaded at: https://www.mdpi.com/article/10.3390/pathogens12060788/s1, File S1: Eighteen mt genomes of rat lungworm used in the data analysis; File S2: Comparison between trees based on mt genomes and individual genes; Table S1: Difference matrix of *cytb* sequences; Table S2: Localities and gene types of samples.

Author Contributions: Conceptualization, S.L.; formal analysis, X.T. and S.L.; data collection and analysis, X.T., S.C. and L.D.; writing—original draft preparation, X.T.; writing—review and editing, Y.Q., H.L. and S.L. All authors have read and agreed to the published version of the manuscript.

Funding: This research was funded by National Key Research and Development Program of China, grant numbers 2021YFC2300800, 2021YFC2300804; and the International Development Research Centre, grant number 108100-001.

Institutional Review Board Statement: Not applicable.

Informed Consent Statement: Not applicable.

Data Availability Statement: The supporting data used in the study are available in the Supplementary Materials.

Acknowledgments: We would like to thank Shenning Lu from National Institute of Parasitic Diseases, China CDC for English editing.

Conflicts of Interest: The authors declare no conflict of interest.

References

1. Lv, S.; Zhang, Y.; Steinmann, P.; Zhou, X.N.; Utzinger, J. Helminth infections of the central nervous system occurring in Southeast Asia and the Far East. *Adv. Parasitol.* **2010**, *72*, 351–408. [PubMed]
2. Wang, Q.P.; Lai, D.H.; Zhu, X.Q.; Chen, X.G.; Lun, Z.R. Human angiostrongyliasis. *Lancet Infect. Dis.* **2008**, *8*, 621–630. [CrossRef] [PubMed]
3. Barratt, J.; Chan, D.; Sandaradura, I.; Malik, R.; Spielman, D.; Lee, R.; Marriott, D.; Harkness, J.; Ellis, J.; Stark, D. *Angiostrongylus cantonensis*: A review of its distribution, molecular biology and clinical significance as a human pathogen. *Parasitology* **2016**, *143*, 1087–1118. [CrossRef]
4. Chen, H.T. A preliminary report on a survey of animal parasites of Canton, China, rats. *Lingnan Sci. J.* **1933**, *12*, 65–74.
5. Nomura, S.; Lin, P.H. First case report of human infection with *Haemostrongylus ratti* Yokogawa. *Taiwan No Ikai* **1945**, *3*, 589–592.
6. Bailey, C.A. *An Epidemic of Eosinophilic Meningitis, a Previously Undescribed Disease, Occuring on Ponape, Eastern Carolines*; Report No. 7; Naval Medical Research Institute, National Naval Medical Center: Bethesda, MD, USA, 1948; pp. 1–23.
7. Rosen, L.; Laigret, J.; Bories, S. Observations on an outbreak of eosinophilic meningitis on Tahiti, French Polynesia. *Am. J. Hyg.* **1961**, *74*, 26–42.
8. Alicata, J.E.; Jindrak, K. *Angiostrongylosis in the Pacific and Southeast Asia*; Thomas Publisher: Springfield, IL, USA, 1970; p. 105.
9. Solorzano Alava, L.F.; Martini Robles, L.; Hernandez Alvarez, H.; Sarracent Perez, J.; Muzzio Aroca, J.; Rojas Rivero, L. *Angiostrongylus cantonensis*: An emerging parasite in Ecuador. *Rev. Cub. Med. Trop.* **2014**, *66*, 20–33.
10. Foronda, P.; Lopez-Gonzalez, M.; Miquel, J.; Torres, J.; Segovia, M.; Abreu-Acosta, N.; Casanova, J.C.; Valladares, B.; Mas-Coma, S.; Bargues, M.D.; et al. Finding of *Parastrongylus cantonensis* (Chen, 1935) in *Rattus rattus* in Tenerife, Canary Islands (Spain). *Acta Trop.* **2010**, *114*, 123–127. [CrossRef] [PubMed]
11. Garcia, M.H.; Moraes, C.; Almada, G.L.; Carvalho, O.S.; Caldeira, R.L.; Mendonça, C.L.; Zanon, D.M.; Antunes, F.A.; Santelli, A.C.; Castro, S.; et al. First reported outbreak of eosinophilic meningitis caused by *Angiostrongylus cantonensis* in Brazil. In Proceedings of the International Conference on Emerging Infectious Diseases, Atlanta, GA, USA, 16–19 March 2008; p. 59.
12. Lowe, S.; Browne, M.; Boudjelas, S.; De Poorter, M. *100 of the World's Worst Invasive Alien Species A selection from the Global Invasive Species Database*; The Invasive Species Specialist Group (ISSG) a specialist group of the Species Survival Commission (SSC) of the World Conservation Union (IUCN): Auckland, New Zealand, 2000; p. 12.
13. Kliks, M.M.; Palumbo, N.E. Eosinophilic meningitis beyond the Pacific Basin: The global dispersal of a peridomestic zoonosis caused by *Angiostrongylus cantonensis*, the nematode lungworm of rats. *Soc. Sci. Med.* **1992**, *34*, 199–212. [CrossRef]

14. Mead, A.R. *The Giant African Snail: A Problem in Economic Malacology*; University of Chicago Press: Chicago, IL, USA, 1961; p. 257.
15. Kliks, M.M.; Kroenke, K.; Hardman, J.M. Eosinophilic radiculomyeloencephalitis: An angiostrongyliasis outbreak in American Samoa related to ingestion of *Achatina fulica* snails. *Am. J. Trop. Med. Hyg.* **1982**, *31*, 1114–1122. [CrossRef]
16. Graeff-Teixeira, C. Expansion of *Achatina fulica* in Brazil and potential increased risk for angiostrongyliasis. *Trans. R. Soc. Trop. Med. Hyg.* **2007**, *101*, 743–744. [CrossRef]
17. Joshi, R.C.; Sebastian, L.S. *Global Advances in Ecology and Management of Golden Apple Snails*; PhilRice: Nueva Ecija, Philippines, 2006; p. 588.
18. Lv, S.; Zhang, Y.; Steinmann, P.; Zhou, X.N. Emerging angiostrongyliasis in Mainland China. *Emerg. Infect. Dis.* **2008**, *14*, 161–164. [CrossRef] [PubMed]
19. Bhaibulaya, M.; Cross, J.H. *Angiostrongylus malaysiensis* (Nematoda: Metastrongylidae), a new species of rat lung-worm from Malaysia. *Southeast Asian J. Trop. Med. Public Health* **1971**, *2*, 527–533. [PubMed]
20. Bhaibulaya, M. A new species of *Angiostrongylus* in an Australian rat, *Rattus fuscipes*. *Parasitology* **1968**, *58*, 789–799. [CrossRef]
21. Prociv, P.; Spratt, D.M.; Carlisle, M.S. Neuro-angiostrongyliasis: Unresolved issues. *Int. J. Parasitol.* **2000**, *30*, 1295–1303. [CrossRef]
22. Stafford, E.E.; Tanudjaja, S.; Purnomo; Carney, W.P. Angiostrongylus malaysiensis in Indonesia. *Southeast Asian J. Trop. Med. Public Health* **1976**, *7*, 490–491.
23. Lim, B.L.; Ramachandran, C.P. Ecological studies on *Angiostrongylus malaysiensis* (Nematoda: Metastrongylidae) in Malaysia. In *Studies on Angiostrongyliasis in Southeast Asia and Australia*; John, H.C., Ed.; U. S. Naval Medical Research Unit No. 2: Taipei, Taiwan, 1979; p. 164.
24. Watthanakulpanich, D.; Jakkul, W.; Chanapromma, C.; Ketboonlue, T.; Dekumyoy, P.; Lv, Z.; Chan, A.H.E.; Thaenkham, U.; Chaisiri, K. Co-occurrence of *Angiostrongylus malaysiensis* and *Angiostrongylus cantonensis* DNA in cerebrospinal fluid: Evidence from human eosinophilic meningitis after ingestion of raw snail dish in Thailand. *Food Waterborne Parasitol.* **2021**, *24*, e00128. [CrossRef]
25. Mallaiyaraj Mahalingam, J.T.; Calvani, N.E.D.; Lee, R.; Malik, R.; Šlapeta, J. Using cerebrospinal fluid to confirm *Angiostrongylus cantonensis* as the cause of canine neuroangiostrongyliasis in Australia where *A. cantonensis* and *Angiostrongylus mackerrasae* co-exist. *Curr. Res. Parasitol. Vector Borne Dis.* **2021**, *1*, 100033. [CrossRef]
26. Valentyne, H.; Spratt, D.M.; Aghazadeh, M.; Jones, M.K.; Šlapeta, J. The mitochondrial genome of Angiostrongylus mackerrasae is distinct from A. cantonensis and A. malaysiensis. *Parasitology* **2020**, *147*, 681–688. [CrossRef]
27. Dumidae, A.; Janthu, P.; Subkrasae, C.; Dekumyoy, P.; Thanwisai, A.; Vitta, A. Genetic characterization of Angiostrongylus larvae and their intermediate host, *Achatina fulica*, in Thailand. *PLoS ONE* **2019**, *14*, e0223257. [CrossRef]
28. Lv, S.; Zhang, Y.; Steinmann, P.; Utzinger, J.; Zhou, X.N. The genetic variation of *Angiostrongylus cantonensis* in the People's Republic of China. *Infect. Dis. Poverty* **2017**, *6*, 125. [CrossRef]
29. Jhan, K.Y.; Cheng, C.J.; Jung, S.M.; Lai, Y.J.; Chen, K.Y.; Wang, L.C. Co-Therapy of Albendazole and Dexamethasone Reduces Pathological Changes in the Cerebral Parenchyma of Th-1 and Th-2 Dominant Mice Heavily Infected with *Angiostrongylus cantonensis*: Histopathological and RNA-seq Analyses. *Biomolecules* **2021**, *11*, 536. [CrossRef]
30. Červená, B.; Modrý, D.; Fecková, B.; Hrazdilová, K.; Foronda, P.; Alonso, A.M.; Lee, R.; Walker, J.; Niebuhr, C.N.; Malik, R.; et al. Low diversity of *Angiostrongylus cantonensis* complete mitochondrial DNA sequences from Australia, Hawaii, French Polynesia and the Canary Islands revealed using whole genome next-generation sequencing. *Parasit Vectors* **2019**, *12*, 241. [CrossRef]
31. Lv, S.; Zhang, Y.; Liu, H.X.; Hu, L.; Yang, K.; Steinmann, P.; Chen, Z.; Wang, L.Y.; Utzinger, J.; Zhou, X.N. Invasive snails and an emerging infectious disease: Results from the first national survey on *Angiostrongylus cantonensis* in China. *PLoS Negl. Trop. Dis.* **2009**, *3*, e368. [CrossRef] [PubMed]
32. Lv, S.; Zhang, Y.; Zhang, L.; Liu, Q.; Liu, H.X.; Hu, L.; Wei, F.R.; Steinmann, P.; Graeff-Teixeira, C.; Zhou, X.N.; et al. The complete mitochondrial genome of the rodent intra-arterial nematodes *Angiostrongylus cantonensis* and *Angiostrongylus costaricensis*. *Parasitol. Res.* **2012**, *111*, 115–123. [CrossRef]
33. Yong, H.S.; Song, S.L.; Eamsobhana, P.; Lim, P.E. Complete mitochondrial genome of *Angiostrongylus malaysiensis* lungworm and molecular phylogeny of Metastrongyloid nematodes. *Acta Trop.* **2016**, *161*, 33–40. [CrossRef]
34. Yong, H.S.; Song, S.L.; Eamsobhana, P.; Goh, S.Y.; Lim, P.E. Complete mitochondrial genome reveals genetic diversity of *Angiostrongylus cantonensis* (Nematoda: Angiostrongylidae). *Acta Trop.* **2015**, *152*, 157–164. [CrossRef] [PubMed]
35. Gasser, R.B.; Jabbar, A.; Mohandas, N.; Schnyder, M.; Deplazes, P.; Littlewood, D.T.; Jex, A.R. Mitochondrial genome of *Angiostrongylus vasorum*: Comparison with congeners and implications for studying the population genetics and epidemiology of this parasite. *Infect. Genet. Evol.* **2012**, *12*, 1884–1891. [CrossRef]
36. Clement, M.; Posada, D.; Crandall, K.A. TCS: A computer program to estimate gene genealogies. *Mol. Ecol.* **2000**, *9*, 1657–1659. [CrossRef] [PubMed]
37. Librado, P.; Rozas, J. DnaSP v5: A software for comprehensive analysis of DNA polymorphism data. *Bioinformatics* **2009**, *25*, 1451–1452. [CrossRef] [PubMed]
38. Alicata, J.E. The presence of *Angiostrongylus cantonensis* in islands of the Indian Ocean and probable role of the giant African snail, *Achatina fulica*, in dispersal of the parasite to the Pacific Islands. *Can. J. Zool.* **1966**, *44*, 1041–1049. [CrossRef] [PubMed]
39. Ramachandran, S.; Deshpande, O.; Roseman, C.C.; Rosenberg, N.A.; Feldman, M.W.; Cavalli-Sforza, L.L. Support from the relationship of genetic and geographic distance in human populations for a serial founder effect originating in Africa. *Proc. Natl. Acad. Sci. USA* **2005**, *102*, 15942–15947. [CrossRef] [PubMed]

40. Eamsobhana, P.; Song, S.L.; Yong, H.S.; Prasartvit, A.; Boonyong, S.; Tungtrongchitr, A. Cytochrome c oxidase subunit I haplotype diversity of *Angiostrongylus cantonensis* (Nematoda: Angiostrongylidae). *Acta Trop.* **2017**, *171*, 141–145. [CrossRef] [PubMed]
41. Bain, R.H.; Hurley, M.M. A biogeographic synthesis of the amphibians and reptiles of Indochina. *Bull. Am. Mus. Nat. Hist.* **2011**, *360*, 1–138. [CrossRef]
42. Qian, M.B.; Chen, Y.D.; Liang, S.; Yang, G.J.; Zhou, X.N. The global epidemiology of clonorchiasis and its relation with cholangiocarcinoma. *Infect. Dis. Poverty* **2012**, *1*, 4. [CrossRef]
43. IARC. A review of human carcinogens. In *IARC Monographs on the Evaluation of Carcinogenic Risks to Humans*; World Health Organization: Geneva, Switzerland, 2012; Volume 100B, p. 441.
44. Lee, J.D.; Chung, L.Y.; Wang, L.C.; Lin, R.J.; Wang, J.J.; Tu, H.P.; Wu, Z.D.; Yen, C.M. Sequence analysis in partial genes of five isolates of *Angiostrongylus cantonensis* from Taiwan and biological comparison in infectivity and pathogenicity between two strains. *Acta Trop.* **2014**, *133*, 26–34. [CrossRef]
45. Peng, J.; He, Z.P.; Zhang, S.; Lun, Z.R.; Wu, Z.D.; Fan, C.K.; Brown, C.L.; Cheng, P.C.; Peng, S.Y.; Yang, T.B. Phylogeography of *Angiostrongylus cantonensis* (Nematoda: Angiostrongylidae) in southern China and some surrounding areas. *PLoS Negl. Trop. Dis.* **2017**, *11*, e0005776. [CrossRef]

pathogens

Article

Barriers with Valve Mechanisms Are Predicted to Protect Crops from Slug Carriers of Rat Lungworm Disease

Genevieve C. Pang *, Amy T. Hou, Ryan Tamashiro, Kristin M. Mills and Lorrin W. Pang

Hawaii State Department of Health, Wailuku, HI 96793, USA; amy.hou.nsw@doh.hawaii.gov (A.T.H.); lorrin.pang@doh.hawaii.gov (L.W.P.)
* Correspondence: genevieve.pang.nsw@doh.hawaii.gov

Abstract: Angiostrongyliasis (Rat Lungworm disease) is an emerging parasitic disease caused by the ingestion of gastropods infected with the neurotropic nematode *Angiostrongylus cantonensis*. The reduction of crop infestation with infected slug carriers may vary widely by protection method. We explored the application of barriers with valve mechanisms, whereby selective directional forces caused a greater number of slugs to exit than enter the protected plot, leading to decreased slug population densities at a steady state. Using field data, we constructed predictive models to estimate slug population densities at a steady state in protected plots with (1) no valve effect, (2) a valve effect, (3) no valve effect with a single breach of the barrier, (4) a valve effect with a single breach of the barrier, (5) a valve effect with a constant breach of the barrier, and (6) a repelling effect. For all scenarios, plots protected using a barrier with a valve effect had consistently lower slug densities at a steady state. Our findings support the use of barriers with valve mechanisms under different conditions, and potentially in combination with other interventions to reduce the contamination of crops by slug carriers of *A. cantonensis*. Improving barriers extends beyond disease mitigation to economic and cultural impacts on the local farmer and consumer communities.

Keywords: Angiostrongyliasis; *Angiostrongylus cantonensis*; *Parmarion martensi*; rat lungworm disease; valve effect; barrier; deterministic model; crops

Citation: Pang, G.C.; Hou, A.T.; Tamashiro, R.; Mills, K.M.; Pang, L.W. Barriers with Valve Mechanisms Are Predicted to Protect Crops from Slug Carriers of Rat Lungworm Disease. *Pathogens* **2023**, *12*, 847. https://doi.org/10.3390/pathogens12060847

Academic Editor: Susan Jarvi

Received: 1 April 2023
Revised: 2 June 2023
Accepted: 15 June 2023
Published: 19 June 2023

1. Introduction

Angiostrongyliasis (rat lungworm disease) is an emerging parasitic disease caused by the neurotropic nematode *Angiostrongylus cantonensis*, which uses gastropods (i.e., snails and slugs) as intermediate hosts and rats as definitive hosts to complete its life cycle. This disease was discovered in southern China in the 1930s and has since spread widely throughout Southeast Asia, Japan, Australia, South America, Southeastern United States, and several island chains, including the Caribbean and Hawaii. The wide distribution of this disease can be attributed, in part, to the proximity of carrier snails and slugs to human habitations and farms, in addition to the rapid and ubiquitous dispersal of rat hosts.

Humans can become accidental hosts when they ingest produce containing uncooked or partially cooked slugs that are infected with juvenile stages of *A. cantonensis*. In humans, larvae die upon reaching the central nervous system, causing a significant inflammatory response that can result in neurologic symptoms and eosinophilic meningitis [1]. Treatment options for this disease are limited, with some evidence for the effectiveness of anthelminthics and corticosteroids [2–4].

Over the past two decades, Hawaii state has experienced an increased incidence of rat lungworm disease. This pathogen has been detected on five of the six most inhabited islands (Oahu, Maui, Hawaii, Kauai, and Lanai), with infection prevalence of gastropods estimated for Kauai, Hawaii, Maui Nui (including Maui, Lanai, and Molokai), and Oahu at 34%, 33%, 18%, and 10%, respectively [5]. This epidemic has been spurred by the documented invasion in 2004 of *Parmarion martensi* [6], a semi-slug that inhabits peridomicile settings, which has since become the primary gastropod carrier of *A. cantonensis* in Hawaii state [7,8]. Although *P. martensi* had been established as the primary carrier species of *A. cantonensis* in Hawaii, a 2014 survey identified a total of 16 carrier gastropod species, some with an infection prevalence approaching 30% [8]. Furthermore, among-island differences in host and pathogen subcommunities likely contribute to observed differences in the carrier species at local scales. For example, in Maui, a recent survey for samples collected from 2016 to 2017 brought the new total of carrier species in Hawaii state to 21, with a higher infection prevalence estimated for *Deroceras reticulatum* (50%) than *P. martensi* (31%) [9]. Therefore, various slug species may contribute to the spread of this disease within Hawaii state, requiring monitoring and control measures of all potential carrier species for a given island. Mitigation strategies include public education efforts on best practices for preparing produce [10], active monitoring of sentinel species (e.g., hind-leg paralysis in juvenile dogs), and the application of barriers or poisoned bait to reduce the number of snails and slugs infesting crops.

Historically, barriers with valve effects have been used to both amplify and reduce the population densities of organisms. A valve influences the direction of animal movement and can increase or decrease the density of animals on either side of the barrier. This directionality is essential to maintain spatial differences in density at a steady state [11]. Barriers to slugs may apply this valve design, whereby slugs can more easily leave than enter a protected area (Table 1). The time to reach a steady state may be affected directly by the size and shape of the internal area and rate of animal movements, and indirectly by population dynamics (e.g., births and deaths) and seasonality (e.g., high or low season) of the external population [12].

Table 1. The effects of valves on the internal population density of organisms. Arrows represent the direction of movement of animals, red lines represent electric barriers, and shaded regions represent the protected plot inside of the slug barrier.

	Valve	No Valve
Type	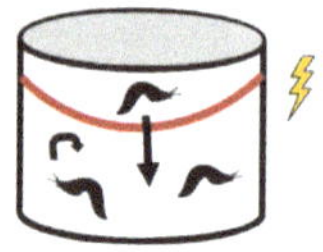	
Direction	Entry < Exit	Entry = Exit
Effect on Internal Density	Decrease	None
Distinguishing Factors	Gravity, Electricity	Electricity

An apparatus with a novel valve design was previously created and tested in a laboratory and field study to combat rat lungworm disease slug carriers [13]. In a laboratory setting, *P. martensi* was observed to readily climb vertical surfaces of a multitude of barrier materials. However, the addition of electrified wires placed on the outside of the vertical fence surrounding a crop created a valve effect; entering slugs were shocked, and either retreated or fell back outside, and slugs inside the protected area that exited over the top of the fence were shocked and fell across the barrier to the outside [13]. This combination of electricity and gravity created a selective directional force such that the overall number of slugs exiting the protected plot was greater than those entering. In the field study, after approximately three weeks, the internal population density of the dominant local species, *D. reticulatum*, was reduced by 90% in the protected plot at a steady state [13]. During the field study, researchers noted an unexpected breach of the barrier, whereby vegetation cuttings served as a convenient bridge over the wall; the resulting spike in the internal slug density returned to a steady state in approximately one week [13]. This observation indicated that *P. martensi* dispersion occurred on a much shorter timescale than reproduction and death, suggesting that birth and death rates may be excluded from future predictive models of valve effects, following the assumptions of movement ecology [14,15].

To explore how barriers with valve effects may reduce slug population densities in protected areas, we constructed a predictive model based on previous invasion models that mechanistically described organismal movement patterns in response to barriers [16]. We used findings from a previous field experiment to estimate the effect of a barrier apparatus with a valve mechanism on slug population densities [13], as well as laboratory experiments from the literature for slug speeds from which to estimate the velocity used in these models [12] (see Appendix A). We investigated the valve mechanism behind previously observed differences between the densities of internal and external slug populations at steady state. We also used these models to address practical questions posed by farmers facing habitat-specific challenges causing regular or irreparable barrier breaches. This model can be used to explore different scenarios, as well as predict how new unforeseen environmental conditions factor into controlling slug populations and reducing human disease.

2. Materials and Methods

To explore the real-world applications of slug barriers with valve mechanisms, we compared the internal and external slug population densities at a steady state under different conditions. We included six possible scenarios for slug control and their effects on the resulting populations at a steady state: (1) no valve effect, (2) a valve effect, (3) no valve effect and single breach of the barrier, (4) a valve effect and single breach of the barrier, (5) a valve effect and various levels of a constant breach of the barrier, and (6) a repelling effect. We built a deterministic model for each scenario using parameter estimates from previous fields and laboratory findings [13] using R programming language [17].

These scenarios are presented in order of complexity, with comparisons made to previous scenarios. Scenarios 1 and 2 investigate the internal population density at a steady state of plots without and with an added valve effect favoring exit over entry, respectively. Scenarios 3 and 4 investigate the resulting internal population density after a temporary breach of barriers without and with an added valve effect. Scenario 5 investigates the effect of a barrier with a valve effect when there is a constant breach of the barrier, allowing slugs to travel between populations on either side of the barrier. Scenario 6 investigates a repelling effect (i.e., barrier materials that reduce slug crossing, such as a zone of salt, diesel oil, copper, or an electric barrier laid horizontally). All scenarios assume the direction of slugs to be random.

In scenarios 1,2,3, and 4, we used the following mathematical equation to determine $y(t)$, the internal slug population as a function of time. We referred to D as the external population density, which was assumed to be constant to allow for relative comparisons of barrier effects with valve mechanisms [13]. The terms P and Q represented the proportion of slugs crossing the barrier (entering and exiting) once they reached it, respectively. This equation contained the constant K, which adjusted the baseline slug population in the protected plot. We defined C as the circumference of the barrier, V as the vector of slug speed and direction, A as the area of the protected plot, and t as time.

$$y(t) = D\left(\frac{P}{Q}\right) + Ke^{-(\frac{CVQ}{A})t} (\text{Scenarios } 1, 2, 3, 4, 6)$$

The P/Q term represented the ratio of the number of slugs entering to exiting the protected plot once they reached the barrier. In future models, this ratio may be modified to affect the strength and direction of the valve effect. In scenario 1 (no valve effect), $P/Q = 1$, as the ratio of slugs entering and exiting via the barrier was equal (Figure 1a). In scenario 2 (valve effect), $P < Q$, as a smaller ratio of slugs entered via the barrier, than exited (Figure 1a). We modified K to explore a single breach scenario without (scenario 3) and with (scenario 4) a valve effect. We explored a single breach at the start of the experiment that made the internal population density twice that of the external population immediately after the breach was repaired, where $K = D$ without a valve effect and $K = 2D - (DP/Q)$ with a valve effect (Figure 1b). In future models, the term K may be modified further to fit any baseline population post-breach.

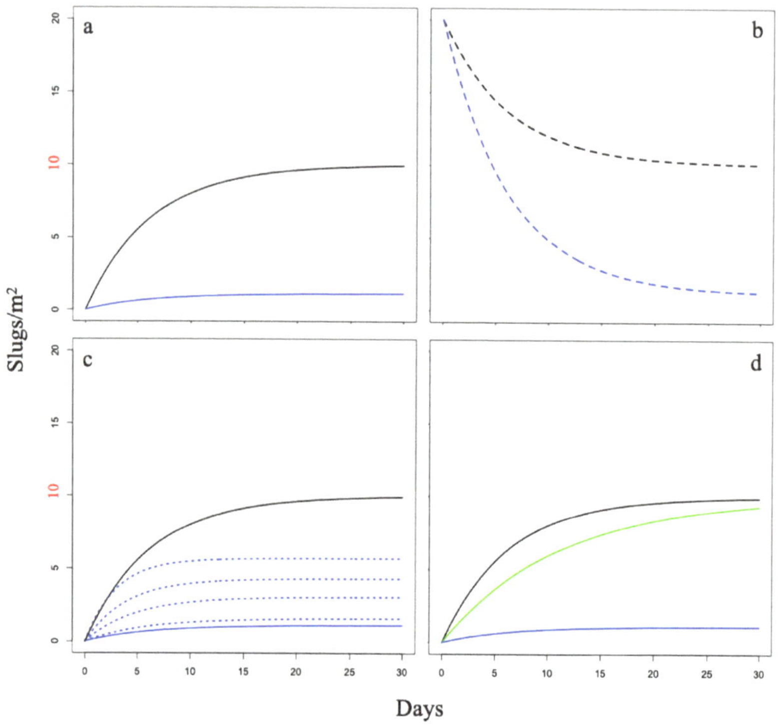

Figure 1. Internal population densities in protected plots in (**a**) scenarios 1 (without a valve effect) (black) and 2 (with a valve effect) (blue). Internal population densities in (**b**) scenarios 3 (without a valve effect) (black) and 4 (with a valve effect) (blue) after a single barrier breach, where the internal density was initially raised to 200% of external density. Internal population densities (**c**) without a valve effect (solid black), with a valve effect (solid blue), and in scenario 5 with four levels (5%, 25%, 50%, and 95% the size of the barrier circumference) of barrier breaches allow a constant flux of slugs between the internal and external populations (dotted). Internal population density (**d**) without a valve effect (black), with a valve effect (blue), and in scenario 6 with barrier materials creating a repelling effect (green) that slows the rate at which the internal population reaches its steady state. For all scenarios, the external population density was set to 10 slugs/m^2 (red).

In scenario 5, we used the following mathematical equation to determine $y(t)$, the internal slug population at a steady state with a constant breach (e.g., an open tunnel or many small tunnels present in porous soil) throughout the experiment, and a valve effect in place (Figure 1c). This scenario was based on a question raised by farmers on the island of Hawaii, where the ground has porous gravel that creates pathways where slugs might tunnel under the barrier. The new term T represented the proportion of slugs crossing (entering and exiting) the tunnel (s) once they reached the perimeter (barrier circumference) of the tunnel; here, T is the same for entry and exit (no valve effect for the tunnel). The term U was the entrance and exit circumference of the tunnels (assumed to be the same). We defined C and T as the circumference of the barrier and barrier breach, respectively, V as the velocity of the slugs [18], A as the area of the protected plot, and t as time.

$$y(t) = D\frac{(CP+UT)}{(CQ+UT)} + Ke^{-(\frac{V}{A})(CQ+UT)t} \text{(Scenario 5)}$$

In scenario 6, we modified P and Q to explore a repelling effect that changes the proportion of slugs that cross the barrier once encountered. To investigate this effect independent of a valve effect, we set $P/Q = 1$ and modified their values.

For all scenarios with valve effects, parameters $P = 0.1$ and $Q = 0.9$; these estimates were based on previous findings in which the ratio of slug densities in the experimental treatment to control plots at a steady state was 9:1, respectively [13]. The experimental data used to estimate the valve effect were taken from two field experiments conducted from May 2020 to February 2021 at a local organic farm in Kula, Hawaii [13]. Infested produce was reported by the farm owner, and surveys confirmed the presence of *D. reticulatum* at this site. Two field experiments were conducted over a period of 10 and 25 weeks, respectively. The first compared slug densities in a plot protected by a barrier with a valve mechanism (electricity) against two control plots protected by barriers without a valve mechanism (no electricity); the second control plot was included to confirm that there was no deterring effect of metal barrier materials on density. The second experiment compared slug densities in a protected and unprotected plot. Plots were approximately 6 m^2 and treated with pellets at the start of each experiment. A barrier breach occurred in the first experiment that was repaired, which resulted in a temporary spike in the treatment plot density followed by a return to previous levels. In all models, slug velocity was set to $V = 0.18$ m/h; this estimate was generated by calculating the velocity (see Appendix A) using the median of an observed range of movement rates for *D. reticulatum* in a laboratory setting [12]. Placeholder values were used for all other parameter estimates at t_0; the same values for these parameters were used in all models to test the relative effects of different scenarios on observed population densities.

3. Results

The lowest internal population density at steady state (as variable time t approached infinity) resulted from a valve effect (scenarios 2 and 4), with and without a single barrier breach, respectively (Figure 1a,b). In both scenarios, the internal population density approached a density of approximately 10% of the external population density. The second lowest internal population density resulted from a valve effect and constant breach (Figure 1c) of the barrier (scenario 5). In this scenario, the internal population density approached approximately 40% of the external population density at a steady state. The highest internal population density resulted from barriers without a valve effect (scenarios 1, 3, and 6), without and with a single barrier breach (Figure 1a,b), and with a repelling effect (Figure 1d), respectively. In these scenarios, the internal population density approached the same density of the external population at a steady state (i.e., the barrier had essentially no effect of reducing the internal population density of slugs); in other words, with no valve effect, it is only a matter of time before the internal and external population densities are the same at steady state.

In addition to predicting outcomes at a steady state, this model also elucidates key parameters that may determine the rate at which a steady state is reached. This model predicts that a steady state will be reached more rapidly at sites where slugs move at greater velocities, as barrier circumference increases, plot areas are made smaller, a greater proportion of slugs exit the protected plot (i.e., a strong valve effect), and travel via tunnels under the barrier increases (i.e., related to larger tunnel circumferences and proportions of slugs crossing).

4. Discussion

This study demonstrates how different barrier designs can lead to potentially very different outcomes for rat lungworm disease mitigation via control of slug carrier populations. We explore the underlying mechanisms affecting population densities of slug carrier species in different scenarios: in response to barriers with and without a valve effect, when barriers are breached and repair is both feasible and infeasible, and barriers with a repelling versus a valve effect.

A key finding of this study is that barriers with valve effects are essential to reduce and maintain lower densities of slugs at a steady state in protected areas. In the absence of a valve effect, the internal population density eventually approached the same value as the external population density; this outcome is predicted to occur regardless of whether the starting density of the internal population is lower (e.g., the internal plot is initially cleared of slugs) or higher (e.g., slugs invaded the internal plot) relative to the external population. With a valve effect, a repaired single breach was predicted to create an initial spike in the internal population density that eventually returned to the same internal density at a steady state as that in the absence of a breach. Left unrepaired, a constant breach reached a steady state density that was higher than when there was no breach (and a valve effect), but lower than when there was no valve effect (and a breach). Thus, an apparatus with a valve effect is predicted to sustainably reduce slug population densities, even when the efficiency of the barrier is reduced by breach events.

While repelling effects may delay the invasion of slugs into a protected area by reducing the rate of crossing a barrier, they do not produce the same outcomes as barriers with valve effects. Rather, barriers that employ only repelling effects do completely prevent the passage of slugs across that barrier [19] and are predicted to simply reduce the rate at which the densities of the protected plot and external populations approach the same value at a steady state (Figure 1d). In theory, barriers without valve effects that severely delay encroachment on a protected crop (e.g., on time scales comparable to a crop's plant-to-harvest cycle) could have practical applications for maintaining lower slug densities prior to harvest. However, installing such barriers that employ only repelling effects would not maintain reduced slug densities at a steady state, resulting in higher densities within protected areas for additional crops planted later in the same growing season. Past barrier designs have commonly focused on the repelling effects of barrier materials, such as copper, to deter slugs [20]. While there has been some evidence that copper can exclude slugs if used in conjunction with a repellent [21,22], the anecdotal evidence for its usefulness for reducing slug densities is mixed [23]. The inconclusive findings of previous investigations of copper could be an artifact of sampling timing due to potentially high variation in density estimates obtained prior to reaching a steady state. However, despite limited evidence for deterrent materials, repelling and valve effects need not be mutually exclusive; plots protected by barriers containing a valve effect with a repelling effect may approach lower steady-state levels inversely proportional to the strength of the valve effect. However, without a valve effect, the internal and external densities are predicted to eventually approach the same value [23], providing support for the use of valve mechanisms for the long-term reduction of slug carriers in protected areas.

Parameters affecting the rate at which steady state is reached are of key interest to predict how rapid target outcomes (e.g., observed reductions in slug densities) may be achieved. We predicted a more rapid approach towards a steady state in systems with slug species that move at greater velocities, large ratios of barrier circumferences to plot areas, and a higher exit rate of slugs from the protected plot. In cases where there is a constant breach of the barrier, increased movements of slugs via tunnels would result in a more rapid approach to steady-state densities. These parameters highlight the importance of considering spatial, temporal, and species-specific factors when designing protective barriers. For example, in scenarios with porous soil and high opportunities for tunneling, slower-growing crops, and dominant slug species that move at faster rates (e.g., *P. martensi*), barriers relying on repelling effects only may be quickly overwhelmed, whereas barriers with valve mechanisms may offer a more effective solution for maintaining lower densities of slugs.

Additionally, factors challenging the model assumptions of homogenous spatial distributions and random movements of slugs within a site may affect the rate at which target outcomes at a steady state are achieved. Potentially critical factors contributing to site-specific, patchy spatial distributions of slugs [24] include density-dependent dispersion and nonlinear movement patterns [12], behavioral interactions [25], and seasonal variation in movement rates. Conditions that reduce encounter rates with the barrier (e.g., reduced slug velocities, repelling conspecific interactions near the barrier, and relatively lower external population densities due to patchiness) are predicted to increase the amount of time to reach a steady state. If a steady state, in which the population density of the protected plot is reduced by 90% relative to the external population, is a target outcome that is time-sensitive, these factors should be considered and tested prior to deploying this apparatus at scale in a given location. In future models, the velocity term may be refined to investigate the potential effects of the above factors on barrier interactions and the resulting rate at which target outcomes are reached. Future field experiments may confirm model predictions using larger plots to investigate the effects of this barrier apparatus at scale.

The simplified deterministic models used in this study did not include birth and death rates or general trends in slug movement direction (i.e., the models assume a random movement of slugs). The parameters for slug reproduction and death were excluded because of the relatively short time scale of slug dispersion compared to its life cycle; this process is supported for other systems in the spatial dynamic population literature, where species distribution occurs on shorter time scales relative to other population dynamics [26]. One generation of *P. martensi* is approximately five to six months [27], whereas, based on previously published field data, the time to reach a steady state is approximately 5 weeks [13]. The parameter estimate for the valve effect in this model was based on experimental field data [13], which accounted for potential competing effects beyond that of the valve effect (e.g., slug preference for higher quality habitat within protected plots). Such opposing effects would contribute to a more conservative estimate of the valve effect modeled in this study. In future models, densities may be modeled over multiple generations or include a modified velocity term (Appendix A) to account for slug habitat preference (e.g., whereby slugs are less likely to exit than enter the protected plot via the barrier).

In future experiments, the model predictions presented in this study could be tested using field experiments that measure changes in slug densities over time in response to breach events, repelling effects, and valve effects. Field and laboratory studies to test the effect of a valve mechanism against other slug species would also improve our ability to accurately predict species-specific outcomes. Additionally, alternative types of valves may improve upon the barrier design explored in this study [13] and could be tested using both field and laboratory experiments. The field experiment demonstrating a 90% reduction in slug densities used relatively small 6 m^2 plots, which are much smaller than large-scale agricultural projects. While the densities of protected plots are predicted to approach a 90% reduction at steady state regardless of plot size, very large-scale plots may contain more nonuniform patches of slug densities and variable strengths of valve effects along the barrier. Additionally, site-specific factors may interact with plot size and barrier circumferences at these larger scales. Some key site-specific factors include weather conditions, slug nutritional state, attractants inside of the plot (e.g., crop type), conspecific interactions (e.g., slug–slug interactions and trails), and slug species. Stochastic models may be useful to account for greater variability in model predictions due to these factors. When the timing of a target outcome is important, in silico experiments would also be useful to determine optimal plot sizes for a given set of site-specific conditions. When possible, the predicted outcomes for different plot sizes should be confirmed prior to the deployment of this barrier apparatus in an agricultural setting.

Practical and effective measures of crop protection against rat lungworm disease gastropod carriers are an essential component to successfully manage the epidemic in Hawaii state, where *A. cantonensis* is broadly distributed and has the potential to expand its range to higher elevations due to warmer average temperatures caused by climate change [8], and annual case counts of this disease are rising [1]. More efficient barriers contributing to the long-term reduction of gastropod carrier densities may not only reduce rat lungworm disease risk, but also has potential applications for a wide array of other diseases transmitted by gastropods to humans (e.g., clonorchiasis, fascioliasis, fasciolopsiasis, opisthorchiasis, paragonimiasis, and schistosomiasis) [28]. Anecdotal observations in field and laboratory settings suggest that this apparatus is effective against snails as well, offering a potentially wide application of these barriers in reducing the densities of various terrestrial gastropods. The addition of valve mechanisms to slug barriers may also help to offset the existing multimillion-dollar costs of terrestrial gastropod-related crop damage in agricultural industries [29]. The apparatus explored in this study [13], which may primarily be employed by farmers, has been designed to be economically attainable at small scales. The cost of materials for the apparatus used in the field component of this project [13] is approximately USD 1944.00 to protect a one-square-hectare plot, including batteries and refugia to monitor changes in the internal slug density of the plot. The cost of materials would likely be reduced if purchased in larger quantities for larger farms, but it also offers a potentially feasible solution for smaller farms. These materials may be reused for future seasons, apart from the batteries. Such solutions not only provide farmers with a means to protect existing crops, but also to potentially grow more delicate crop species that are less resistant to slug herbivory (e.g., napa cabbage). Additionally, for organic farms, this apparatus provides a chemical-free solution to reduce pests' damage to produce.

In conclusion, our findings support the use of valve mechanisms in barriers to rat lungworm disease slug carriers. This key design component is predicted to yield a long-term reduction in the population densities of slug carriers in protected areas. The use of effective protective barriers is essential, not only to mitigate disease risk, but to promote the economic and cultural welfare of farmers and local communities.

Author Contributions: Conceptualization, methods, visualization, interpretation of analyses, and writing, G.C.P., A.T.H., R.T., K.M.M. and L.W.P.; mathematical models, G.C.P., R.T. and L.W.P.; appendix, G.C.P., A.T.H., R.T. and L.W.P. All authors have read and agreed to the published version of the manuscript.

Funding: This research received no external funding.

Institutional Review Board Statement: Not applicable.

Acknowledgments: The authors would like to thank Kim Pham and Nicole Ferguson for their helpful comments and suggestions in the initial development of this project and discussion of its findings.

Conflicts of Interest: The authors declare no conflict of interest.

Appendix A

A more detailed explanation of the mathematical expression will be given here. We will not cover how the solution (integration) for variable Y is derived from the differential equation, which can be found in standard references for ordinary differential equations (ODE). Furthermore, we will not "double-check" to show that this solution for Y does, in fact, fit the ODE. Instead, this appendix will describe in more detail the parameters of the math expression in terms of the physical slug and garden setting. Notably, while the basic equation tracks the number of slugs (abundance), we would like an expression for slug densities (number per unit area). As we describe parameters and how they fit into the equations, it is very useful to keep track and be cognizant of the units to provide an intuitive sense. For example, exponential powers should be in terms of pure numbers (except for the scalar of time) if the independent variable x were to represent time.

The approach to examine the mathematics of the model is to first set up an intuitive tally balancing the change (difference over time, differential) in the number of slugs in the garden. Over an interval of time, this change will be the number entering minus the number leaving. Assume that movement dominates the changing numbers (such as birth, deaths, and predation). Then, we examine the parameters of the equation (the physical aspects of slug movement in our setting) to express the change in slug density rather than numbers. We isolate terms that are time-dependent and those that are not. Based on the type of differential equation, we can solve for the value of density itself. This solution will incorporate the initial parameters and introduce one new parameter, the initial baseline density.

Based on our previously published laboratory and field experiments (13) with this slug/snail barrier, we start with a general equation #1, which balances the inward and outward movements of the number of slugs (abundance). From this first expression, we will determine a solution as shown in equation #2, as well as look at two separate circumstances added onto equation #1: a breach that is fixed after a fixed number of slugs enter and exit through, as shown in equation #3, and a breach which cannot be fixed but allows slugs to enter and exit at a constant rate as shown in equation #4.

Equation #1 (Differential Equation for Slug Abundance)

The change in the number of slugs per unit of time is equal to the number entering (across the barrier) minus the number exiting. There are no alternate pathways (breaches).

$$\frac{dN}{dt} = K_1 - K_2$$

Where: N is the total number (or abundance) of slugs in the garden (number).

Where:

$$K_1 = (C)(V)(pOut)(ProEnt)$$

Where:

C is the circumference of the garden (cm).

V is the slug velocity vector of both speed and direction to reach the barrier (cm per time).

pOut is outside slug density (number per cm^2).

ProEnt is the proportion of slugs that enter across the barrier once they have reached it from the outside.

Where:

$$K_2 = (C)(V)(pIn)(ProEx)$$

Where:

C and V are defined as above.

pIn is the inside slug density, dependent on time (number per cm^2).

ProEx is the proportion of slugs that exit across the barrier once they have reached it from the inside.

Notes:

1. The outside slug density is assumed to be constant, whereas the inside density depends on time (the time from when the barrier is activated with some initial baseline population). This initial baseline population will not appear in the differential equation and will only appear later (after integration) when one solves N itself. Mathematically, it will be a constant of integration that can be set to match the starting baseline internal slug population;

2. It was reported in the laboratory experiments of the original publication (13) that for a vertical barrier, as slugs approached from the top (exiting the garden barrier), they could retreat (crawl back up) while others "crossed" the barrier, either falling over it or crawled across it (very rare). When slugs approached from the bottom (entering), none climbed across the barrier; they either crawled back down or fell back due to the electric shock. Equation #1 allows for two different proportions for entry and exit, and "crossing" the barrier can occur by falling or crawling over it. With the barrier set up outside the garden, the falling movements favor exiting over the entry;

3. Here is a detailed explanation for the slug velocity V; see Figure A1 The velocity vector has two components: speed (*Sp*, which is the magnitude) and direction (θ). The total circumference can be divided into smaller lengths (approaching infinitely small segments as we apply calculus to tally the total number of slugs reaching the entire circumference). Corresponding to each segment, there is an area of flux, defined as the area in which all

slugs have the potential to reach the barrier in time t. The distance of the upper border of the flux area is set to the distance calculated by a slug moving at speed Sp for a time t. This flux area excludes slugs that are too far away. In this flux area, let all the slugs have an average speed, Sp.

To get a notion of the importance of direction (θ) to reach the barrier, at this distance, only slugs traveling in the direction perpendicular to the circumference segment will reach the barrier. At the extreme limit of the distance, only 1/360 will be going in the right direction to reach the barrier. For the slugs at the barrier, half of them will have the correct direction (180/360) to reach it, while the other half will move away.

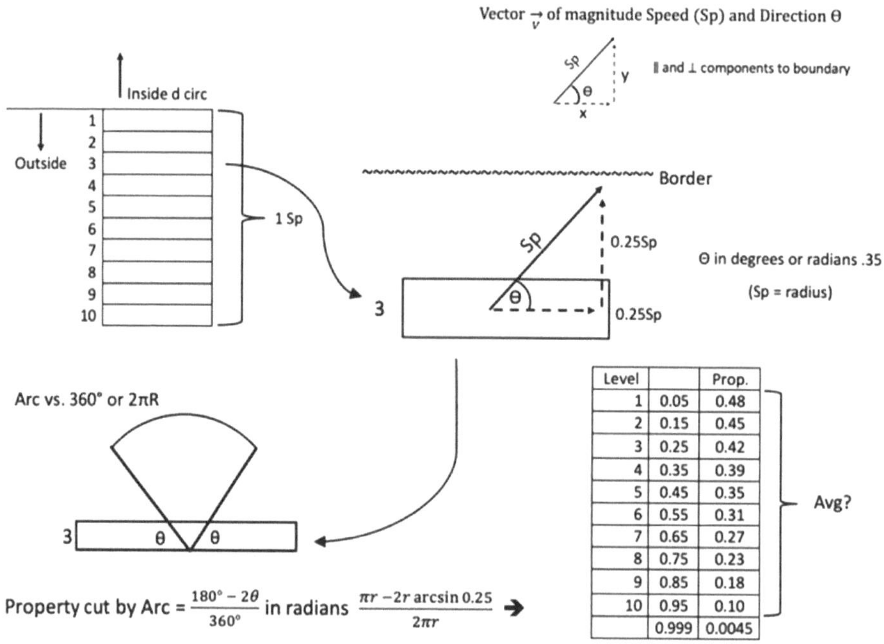

Figure A1. Calculation of the proportion of slugs approaching a barrier based on their initial distance from the barrier and velocity (i.e., speed and direction).

Figure A1 shows a typical flux area along the circumference where slugs enter the garden approaching it from the outside. A similar diagram can be drawn in reverse, where slugs exit the garden by approaching it from the inside. The flux area is broken up into 10 levels, each with midpoint perpendicular distances corresponding to Sp from closest to farthest: 0.05 Sp, 0.15 Sp, 0.25 Sp, . . . , 0.95 Sp. For example, looking at level 3, only slugs with enough perpendicular direction will reach the border at 0.25 Sp; based on trigonometry, the limit will be set by the angle θ, whose sine is 0.25. All slugs traveling more "vertically" than θ off the horizontal axis will reach the border; therefore, the proportion of slugs moving in the right direction will equal the following: $(180 - 2\theta)/360 = 42\%$. The figure shows another column with the proportion of slugs moving in the right direction for each stratum. Across all strata of the flux area, the average of the slugs moving in the right direction is 32%. Alternatively, mathematically one could get an exact proportion by integrating across infinitely small layers with their corresponding proportions bounded by these angles (it is good to convert the angles in radians to avoid switching positive and negative values of sine as one crosses quadrants).

If slug motion is random in all directions, for any given average speed of slugs, only about 32% of the slugs are moving in the right direction to reach the barrier. Where Sp is the average speed of the slug, and its unit is cm/time unit. The unit of time is a scalar to be chosen for the entire equation (i.e., seconds, minutes, hours, days, weeks, etc.). We assume the random direction (θ) of the slug's movement for entering and exiting the garden. For the same value *Sp*, refinement to the model can be made so that if there is an attractant in the garden, there will be a non-random, higher percentage of slugs moving in the direction to reach the barrier from the outside and the opposite effect for the slugs already inside the garden;

4. The number of slugs in the garden N is the fixed area of the garden (A) multiplied by the inside slug density (pIn). The differential equation #1 can be rewritten to convert N into density and to isolate the time-dependent variable, which is the inside slug density pIn.

Where:

$$\frac{dpIn}{dt} = \frac{K_1}{A} - \frac{K_{2'}}{A}(pIn)$$

Where: A is the area of the garden (cm^2).

From the form of the above differential equation of the time-dependent density, the solution for this differential would be:

$$K_2{}' = K_2/pIn$$

$$pIn = (K_1/A)/(K_2{}'/A) + C3 \times e^{-(K_2{}'/A)t}$$

The new constant of integration $C3$ is introduced and can be used to "set" the initial baseline pIn levels (when $t = 0$). Its units are the same as the density and number per cm^2.

Equation #2

To better appreciate the parameters of the mathematical model and to track the unit, let us substitute the parameters for the dummy variables K_1, K_2, and $K_2{}'$, as demonstrated here:

$$\begin{aligned} pIn &= \frac{\left(\frac{(C)(Sp)(0.32)(ProEnt)(pOut)}{A}\right)}{\left(\frac{(C)(Sp)(0.32)(ProEx)}{A}\right)} + C3e^{\frac{-(C)(Sp)(0.32)(ProEx)t}{A}} \\ &= \frac{(ProEnt)(pOut)}{ProEx} + C3e^{\frac{-(C)(Sp)(0.32)(ProEx)t}{A}} \end{aligned}$$

At the steady state when t gets very large, the second term approaches zero, and $pIn/pOut = ProEnt/ProEx$. For example, when it is 10 times easier to exit than to enter, the internal slug density will be one-tenth of the outside density. This difference between the proportion that enters versus exits is what we term the "valve" effect. It is similar to the valve effect of funnel fish traps, except that the funnel is set to concentrate fish inside the trap rather than outside. Alternatively, at a steady state, the internal and external densities will be equal if the chances of entering and exiting are the same. This is true even for very effective barriers without a valve effect, for example, the same electric-barrier system (blocking 90% of slugs) but laid horizontally without the valve effect of shock + gravity, or even barriers with repelling effects, such as copper.

Next, the equation shows the factors which affect how quickly the steady state will be reached. Looking at the components of the factors of the negative exponent, we see that an increased garden circumference, greater speed of slug movement, a high proportion of slugs exiting, and a decreased garden area will approach a steady state faster. Furthermore, the exponent is a pure number with the time scale matching the time units of the slug speed.

Finally, the value of $C3$ is determined algebraically based on whatever we choose as the baseline population density in the garden (when $t = 0$).

As an example of the above principles, suppose we do not attempt any baseline "one-time" clean out of the slugs in the garden. This population will then approach a steady-state level depending on the valve effect (regardless of the intrinsic repelling effect of the material). How quickly this occurs depends on the speed of the slugs, the circumference, and the area of the garden.

Equation #3: Effects of a One-Time Breach

Building upon equations #1 and# 2 (using baseline $pIn = 0$, and $ProEx/ProEnt = 9{:}1$ valve effect, same slug average speed and garden *C* and *A*), the differential equation of slug abundance during a one-time breach will be the following which addresses the entry and exit via the one-time breach:

$$dN/dt = (K_1 + B_{ent}) - (K_2 + B_{ex})$$

Where:

B_{ent} is the number of slugs entering the garden via the breach over time (number per time unit)

B_{ex} is the number of slugs exiting the garden via the breach over time (number per time unit)

The conversion of the equation to track slug densities is similar to what was done for equation #1 but introduces the effects of the breach. Immediately after the breach is fixed, we determine the final endpoint for the internal density, *pIn*. Moving forward after the breach is fixed, this endpoint becomes the new starting point that follows the rules of equation #2, but now with a new non-zero starting point. This principle of using equation #2 but with a new starting point is very useful since it does not really depend on what the breach is. For example, it might not be a pass-through through which slugs enter and exit, but simply a one-time breach of slugs, say slug-infested compost dumped into the garden. Thus, the method of handling slug densities after the breach has ended is nothing more than redefining a baseline slug density and proceeding with equation #2. Additionally, as we showed previously, this initial starting need not be zero and will approach the steady state through the dynamics of the model depending on various parameters. This starting point moving forward after the breach will determine the value C3.

For readers who laboriously tallied parameters during the breach to reach the endpoint density, your work will not go unrewarded. The next situation is a permanent, unrepaired breach, which is nothing more than a description of the model during the breach before it was repaired.

Equation #4: Effects of a Constant Breach

Finally, we address the concern of an "irreparable" tunneling network of rocky, porous ground through which slugs move. Mathematical models should be specific to the nature of the breach. However, if breaches can be quickly repaired (i.e., in our field experiment, the breaches were detected and repaired within 3 days), we have the luxury of moving forward with a no-breach model, with the residual breach effect incorporated by a new baseline starting point.

To highlight the principles of a continuous breach without oversimplification, we assume a similar setting with similar parameters as that described in equations #1 and #2: garden circumference, garden area, slug speed, the random direction of movement, valve effect (9:1), baseline *pIn* (zero). We add in the tunneling through the ground as our constant breach. To keep things simple, passing through the tunnel will follow the rules of passing through the barrier. The tunnel entrance(s) must be approached according to the rules of slug density, flux, and velocity. Of the slugs that reach the tunnel circumferences, there is a proportion that will pass through in either direction (entering or exiting). Assuming there is no valve effect on the tunnel, of those slugs that approach the tunnel entrance(s), the proportion of those entering is the same as those exiting. Furthermore, the tunnel option is treated as an alternative to the barrier: slugs will either choose to pass through the barrier or tunnel, not both.

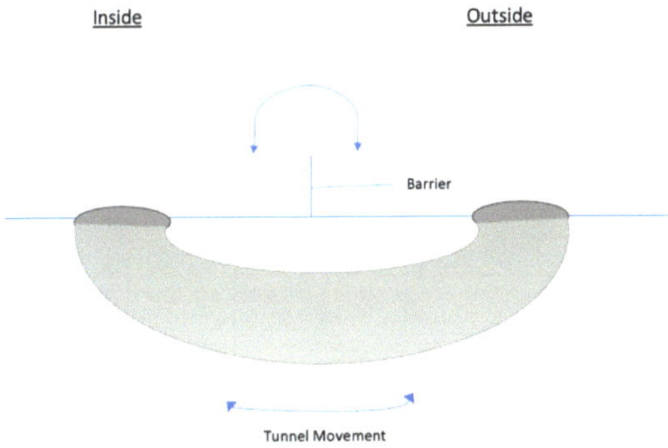

Figure A2. Diagram of the tunnel (constant breach) in which slugs can move between the inside and outside of a plot protected by a barrier.

The assumptions for the barrier are that the internal population density is 0 at baseline, the valve effect is 9:1, and there are no repaired breaches. The assumptions for the tunnel are that internal population density is 0 at baseline, there is no valve effect, and the tunnel effect is proportional to the circumference of the tunnel opening as opposed to its volume. The cumulative effect of multiple tunnels can be represented in a single term where circumferences of multiple tunnels are added together.

The differential equation is the net change in the slug abundance: the number entering minus the number exiting. Slugs can pass (enter and exit) through the tunnel or through the barrier. It is similar to equation #1: $dN/dt = K_1 - K_2$, but now the entry rate K_1 and exit rate K_2 are through the barrier or the tunnel.

$$dN/dt = (K_{1B} + K_{1T}) - (K_{2B} + K_{2T})$$

Where:
K_{1B} is the rate of slugs entering via barrier (number per unit of time).
K_{1T} is the rate of slugs entering via tunnel (number per unit of time).
K_{2B} is the rate of slugs exiting via barrier (number per unit of time).
K_{2T} is the rate of slugs exiting via tunnel (number per unit of time).

The following equation is similar to equation #2, where we converted it to internal density and isolated the parameter of pIn because it is time-dependent:

$$\frac{dpIn}{dt} = \frac{(K_{1B} + K_{1T})}{A} - \frac{(K_{2B'} + K_{2T'})(pIn)}{A}$$

The solution for density at a steady state will be:

$$pIn = \frac{\left(\frac{(K_{1B}+K_{1T})}{A}\right)}{\left(\frac{(K_{2B'}+K_{2T'})}{A}\right)}$$

$$K_{2B'} + K_{2T'} = \frac{(K_{2B} + K_{2T})}{pIn}$$

For variables $K_{2b'}$ and K_{2T}, please refer to the previous definition of $K_{2'}$.
Filling out the parameters for the values of K, we have:

$$\begin{aligned} K_{1T} &= (CT)(Flux)(pOut)(ProT) \\ &= (CT)(V)(pOut)(ProT) \\ &= (CT)(Sp)(0.32)(pOut)(ProT) \end{aligned}$$

Where:
CT is the tunnel circumference (equal on both tunnel ends).
$ProT$ is the proportion of slugs crossing via the tunnel.
Similarly,

$$K_{2T} = (CT)(Sp)(0.32)(pIn)(ProT)$$

Converting to the internal slug density,

$$\frac{dpIn}{dt} = \frac{1}{A}(Total\ rate\ enter - Total\ rate\ exit)$$

With the isolation of variable pIn,

$$\frac{dpIn}{dt} = C_1 - C_2(pIn)$$

Where: C_1 and C_2 are placeholder variables:

$$C_1 = \frac{1}{A}(pOut)(Sp)(0.32)((C)(ProEnt) + (CT)(ProT))$$

$$C_2 = \frac{1}{A}(Sp)(0.32)((C)(ProEx) + (CT)(ProT))$$

Notice that the units of C_1 and C_2 differ by density unit. The unit of C_1 is number/(cm^2 $\times$ unit time). The unit of C_2 is 1/unit time.

Solving for pIn,

$$pIn = \frac{C_1}{C_2} + C3e^{-C_2 t}$$

At a steady state, the second term of the right side of the above equation approaches 0, leading to the following equation:

$$\frac{pIn}{pOut} = \frac{\left(\frac{1}{A}\right)(Sp)(0.32)(Z_1)}{\left(\frac{1}{A}\right)(Sp)(0.32)(Z_2)}$$

Where: Z_1 and Z_2 are placeholder variables such that:

$$Z_1 = (C)(ProEnt) + (CT)(ProT)$$
$$Z_2 = (C)(ProEx) + (CT)(ProT)$$

Z_1/Z_2 is the ratio of the inside density to the outside density at a steady state. This ratio has the form:

$$\frac{a+c}{b+c}$$

When c is large (i.e., a very large tunnel effect) with respect to a and b, the internal population density approaches the external population density at a steady state (in other words, as if there is no barrier).

When c is small (i.e., a very small tunnel effect) with respect to a and b, the internal population density approaches the population density with a valve effect at a steady state (in other words, as if there is no tunnel).

To explore how quickly a steady state is reached with a tunnel effect, investigate the exponent term above. Equilibrium is predicted to be approached faster with increasing speed, barrier circumference, tunnel circumference, the proportion of slugs exiting via the barrier, the proportion of slugs exiting via the tunnel, and decreasing garden area.

References

1. Cowie, R.H.; Ansdell, V.; Dunavan, C.P.; Rollins, R.L. Neuroangiostrongyliasis: Global spread of an emerging tropical disease. *Am. J. Trop. Med. Hyg.* **2022**, *107*, 1166–1172. [CrossRef] [PubMed]
2. Hwang, K.P.; Chen, E.R. Clinical studies on *Angiostrongyliasis cantonensis* among children in Taiwan. *Southeast Asian J. Trop. Med. Public Health* **1991**, *22*, 194–199. [PubMed]
3. Jitpimolmard, S.; Sawanyawisuth, K.; Morakote, N.; Vejjajiva, A.; Puntumetakul, M.; Sanchaisuriya, K.; Tassaneeyakul, W.; Tassaneeyakul, W.; Korwanich, N. Albendazole therapy for eosinophilic meningitis caused by *Angiostrongylus cantonensis*. *Parasitol. Res.* **2007**, *100*, 1293–1296. [CrossRef] [PubMed]
4. Ansdell, V.; Kramer, K.J.; McMillan, J.K.; Gosnell, W.L.; Murphy, G.S.; Meyer, B.C.; Blalock, E.U.; Yates, J.; Lteif, L.; Smith, O.A.; et al. Guidelines for the diagnosis and treatment of neuroangiostrongyliasis: Updated recommendations. *Parasitology* **2021**, *148*, 227–233. [CrossRef] [PubMed]
5. Kim, J.R.; Wong, T.W.; Curry, P.A.; Hayes, K.A.; Cowie, R.H. Modelling the distribution in Hawaii of *Angiostrongylus cantonensis* (rat lungworm) in its gastropod hosts. *Parasitology* **2019**, *146*, 42–49. [CrossRef] [PubMed]
6. Hollingsworth, R.G.; Howe, K.; Jarvi, S.I. Control measures for slug and snail hosts of *Angiostrongylus cantonensis*, with special reference to the semi-slug *Parmarion martensi*. *Hawai'i J. Med. Public Health* **2013**, *72* (Suppl. 2), 75–80.
7. Qvarnstrom, Y.; Bishop, H.S.; da Silva, A.J. Detection of rat lungworm in intermediate, definitive, and paratenic hosts obtained from environmental sources. *Hawai'i J. Med. Public Health* **2013**, *72* (Suppl. 2), 63–69.
8. Kim, J.R.; Hayes, K.A.; Yeung, N.W.; Cowie, R.H. Diverse gastropod hosts of Angiostrongylus cantonensis, the rat lungworm, globally and with a focus on the Hawaiian Islands. *PLoS ONE* **2014**, *9*, e94969. [CrossRef]
9. Yeung, N.W.; Kim, J.R.; Hayes, K.A. Rat lungworm (*Angiostrongylus cantonensis*) in Hawai'i: Updated host gastropod records and distributions on Maui. *Bish. Mus. Occas. Pap.* **2018**, *123*, 19–24.
10. Cowie, R.H. Pathways for transmission of angiostrongyliasis and the risk of disease associated with them. *Hawai'i J. Med. Public Health* **2013**, *72* (Suppl. S2), 70–74.
11. Fogarty, M.J.; Addison, J.T. Modelling capture processes in individual traps: Entry, escapement, and soak time. *ICES J. Mar. Sci.* **1997**, *54*, 193–205. [CrossRef]
12. Ellis, J.; Petrovskaya, N.; Forbes, E.; Walters, K.F.A.; Petrovski, S. Movement patterns of the grey field slug (*Deroceras reticulatum*) in an arable field. *Sci. Rep.* **2020**, *10*, 17970. [CrossRef]
13. Pang, L.; Coppolo, C.; Hauptman, S. An effective barrier to prevent crop contamination by slug vectors of *Angiostrongylus cantonensis*. *Am. J. Trop. Med. Hyg.* **2022**, *106*, 1678–1683. [CrossRef] [PubMed]
14. Arditi, R.; Tyutyunov, Y.; Morgulis, A.; Govorukhin, V.; Senina, I. Directed movement of predators and the emergence of density-dependence in predator-prey models. *Theor. Popul. Biol.* **2001**, *59*, 207–221. [CrossRef] [PubMed]
15. Shaw, A.K.; White, L.A.; Michalska-Smith, M.; Borer, E.T.; Craft, M.E.; Seabloom, E.W.; Snell-Rood, E.C.; Travisano, M. Lessons from movement ecology for the return to work: Modeling contacts and the spread of COVID-19. *PLoS ONE* **2021**, *16*, e0242955. [CrossRef] [PubMed]
16. Azimzade, Y. Invasion front dynamics of interactive populations in environments with barriers. *Sci. Rep.* **2022**, *12*, 826. [CrossRef]
17. R Core Team. R: A language and environment for statistical computing. In *R Foundation for Statistical Computing*; R Core Team: Vienna, Austria, 2022. Available online: https://www.R-project.org/ (accessed on 20 December 2022).
18. Lai, J.H.; del Alamo, J.C.; Rodriguez-Rodriguez, J.; Lasheras, J.C. The mechanics of the adhesive locomotion of terrestrial gastropods. *J. Exp. Biol.* **2010**, *213*, 3920–3933. [CrossRef]
19. Symondson, W.O. Chemical confinement of slugs: An alternative to electric fences. *J. Molluscan Stud.* **1993**, *59*, 259–261. [CrossRef]
20. Schuder, I.; Port, G.; Bennison, J. Barriers, repellents and antifeedants for slug and snail control. *Crop Prot.* **2003**, *22*, 1033–1038. [CrossRef]
21. Turke, M.; Heinze, E.; Andreas, K. Seed consumption and dispersal of ant-dispersed plants by slugs. *Oecologia* **2010**, *163*, 681–693. [CrossRef] [PubMed]
22. Kheirodin, A.; Damavandian, M.R.; Sarailoo, M.H. Mineral oil as a repellent in comparison with other control methods for citrus brown snail, *Caucasotachea lencoranea*. *Afr. J. Agric. Res.* **2012**, *7*, 5701–5707.
23. Watz, J.; Nyqvist, D. Artificial barriers against arionid slug movement. *Crop Prot.* **2021**, *142*, 105525. [CrossRef]
24. Forbes, E.; Back, M.A.; Brooks, A.; Petrovskaya, N.B.; Petrovskii, S.V.; Pope, T.W.; Walters, K.F.A. Locomotor behavrio promotes stability of the patchy distribution of slugs in arable fields: Tracking the movement of individual *Deroceras reticulatum*. *Pest Manag. Sci.* **2020**, *76*, 2944–2952. [CrossRef] [PubMed]
25. Petrovskii, S.; Ellis, J.; Forbes, E.; Petrovskaya, N.; Walters, K.F.A. A predictive model and a field study on heterogenous slug distribution in arable fields arising from density dependent movement. *Sci. Rep.* **2022**, *12*, 2274. [CrossRef] [PubMed]
26. Potts, J.R.; Lewis, M.A. Spatial memory and taxis-driven pattern formation in model ecosystems. *Bull. Math. Biol.* **2019**, *81*, 2725–2747. [CrossRef] [PubMed]
27. Hamilton, L.J.; Tagami, Y.; Kaluna, L.; Jacob, J.; Jarvi, S.I.; Follett, P. Demographics of the semi-slug *Parmarion martensi*, an intermediate host for *Angiostrongylus cantonensis* in Hawai'i, during laboratory rearing. *Parasitology* **2021**, *148*, 153–158. [CrossRef] [PubMed]

28. Lu, X.; Gu, Q.; Limpanont, Y.; Song, L.; Wu, Z.; Okanurak, K.; Lv, Z. Snail-borne parasitic diseases: An update on global epidemiological distribution, transmission interruption and control methods. *Infect. Dis. Poverty* **2018**, *7*, 28. [CrossRef]
29. Jiang, X.; Zheng, P.; Soto, I.; Haubrock, P.J.; Chen, J.; Ji, L. Global economic costs and knowledge gaps of invasive gastropods. *Ecol. Indic.* **2022**, *145*, 109614. [CrossRef]

pathogens

MDPI

Article

Rapid Single-Step Immunochromatographic Assay for *Angiostrongylus cantonensis* Specific Antigen Detection

Praphathip Eamsobhana [1,*], Anchalee Tungtrongchitr [1,*], Darawan Wanachiwanawin [1], Sudarat Boonyong [1] and Hoi-Sen Yong [2]

1 Department of Parasitology, Faculty of Medicine Siriraj Hospital, Mahidol University, Bangkok 10700, Thailand; darawan.wan@mahidol.ac.th (D.W.); sudarat.boo@mahidol.ac.th (S.B.)
2 Institute of Biological Sciences, Faculty of Science, Universiti Malaya, Kuala Lumpur 50603, Malaysia; yong@um.edu.my
* Correspondence: praphathip.eam@mahidol.ac.th (P.E.); anchalee.tun@mahidol.ac.th (A.T.)

Abstract: *Angiostrongylus cantonensis* is the major etiological nematode parasite causing eosinophilic meningitis and/or eosinophilic meningoencephalitis in humans. The rapid global spread of *Angiostrongylus cantonensis* and the emerging occurrence of the infection have exposed the shortcomings of traditional/conventional diagnostics. This has spurred efforts to develop faster, simpler and more scalable platforms that can be decentralized for point-of-need laboratory testing. By far, the point-of-care immunoassays such as the lateral flow assay (LFA) are the best-placed. In this work, a LFA in the form of an immunochromatographic test device (designated *Ac*AgQuick^{Dx}), based on the detection of a circulating *Angiostrongylus cantonensis*-derived antigen, was established using anti-31 kDa *Angiostrongylus cantonensis* antibody as the capture reagent and anti-*Angiostrongylus cantonensis* polyclonal antibody as the indicator reagent. The *Ac*AgQuick^{Dx} was evaluated for its diagnostic potential with a total of 20 cerebrospinal fluids (CSF) and 105 serum samples from patients with angiostrongyliasis and other clinically related parasitic diseases, as well as serum samples from normal healthy subjects. Three of the ten CSF samples from serologically confirmed angiostrongyliasis cases and two of the five suspected cases with negative anti-*Angiostrongylus cantonensis* antibodies showed a positive *Ac*AgQuick^{Dx} reaction. Likewise, the *Ac*AgQuick^{Dx} was able to detect *Angiostrongylus cantonensis* specific antigens in four serum samples of the 27 serologically confirmed angiostrongyliasis cases. No positive reaction by *Ac*AgQuick^{Dx} was observed in any of the CSF (n = 5) and serum (n = 43) samples with other parasitic infections, or the normal healthy controls (n = 35). The *Ac*AgQuick^{Dx} enabled the rapid detection of active/acute *Angiostrongylus cantonensis* infection. It is easy to use, can be transported at room temperature and does not require refrigeration for long-term stability over a wide range of climate. It can supplement existing diagnostic tests for neuroangiostrongyliasis under clinical or field environments, particularly in remote and resource-poor areas.

Keywords: *Angiostrongylus cantonensis*; lateral flow assay; immunochromatographic test; antigen detection; 31-kDa antigen; active angiostrongyliasis

check for updates

Citation: Eamsobhana, P.; Tungtrongchitr, A.; Wanachiwanawin, D.; Boonyong, S.; Yong, H.-S. Rapid Single-Step Immunochromatographic Assay for *Angiostrongylus cantonensis* Specific Antigen Detection. *Pathogens* **2023**, *12*, 762. https://doi.org/10.3390/pathogens12060762

Academic Editor: Timothy G. Geary

Received: 28 March 2023
Revised: 3 May 2023
Accepted: 18 May 2023
Published: 25 May 2023

1. Introduction

Angiostrongylus cantonensis, also known as rat lungworm, is an angiostrongylid nematode parasite that typically resides in the pulmonary arteries and right ventricle of rodents. It is the commonest etiological parasite causing eosinophilic meningitis and/or eosinophilic meningoencephalitis in humans [1,2]. This neurotropic parasite is of increasing public health importance as its geographic distribution now covers a wide part of the world, and new locality records continue to be reported [2–6].

Angiostrongylus cantonensis has a relatively simple heteroxenous life cycle, typically involving a definitive rodent host and a mollusk intermediate host, but it may also use various paratenic hosts [2,7]. Humans are an accidental host, acquiring the infection via the ingestion of raw or poorly cooked snail meat and a variety of paratenic hosts which harbor the third-stage larvae, or green vegetables contaminated with the infective larvae [1–3,6,8]. The most frequently reported symptoms are headache, neck stiffness, paresthesia, fever, visual disturbances, vomiting and nausea [1–5,9]. The migration of the larvae to the brain tissue causes serious central nervous system (CNS) damage, which can result in coma and death of the patient [10,11].

The definitive diagnosis for human neural angiostrongyliasis is based on the detection of immature worms in the cerebrospinal fluid (CSF) or from the eye of infected patients, but such findings are of rare occurrence [9]. The disease is presumptively diagnosed based on presenting symptoms, medical histories, eosinophilic pleocytosis in the CSF, and immunological and/or molecular markers [2–4,12].

Antibody-based immunodiagnostic tests using purified antigens for neuroangiostrongyliasis have been available for decades. Most of the work has centered on the *Angiostrongylus cantonensis* 31-kDa glycoprotein antigen [13,14], 29-kDa antigen [15], and 32-kDa protein [16]. The 31-kDa glycoprotein antigen of *Angiostrongylus cantonensis* has mostly been used in traditional immunoblotting as a diagnostic marker for differential diagnosis in human angiostrongyliasis, with very high sensitivity and specificity [17–19]. On the other hand, the detection of circulating *Angiostrongylus cantonensis* antigens is an alternative immunodiagnostic test to identify the active/acute/early stages of neuroangiostrongyliasis, and it is very much required for the patient management/treatment.

Furthermore, a molecular approach that targets gene sequences of *Angiostrongylus cantonensis* can also assist in an early etiologic diagnosis. Various DNA-based diagnostic techniques that rely on polymerase chain reaction (PCR) to amplify and detect specific *Angiostrongylus cantonensis* DNA molecules have been successfully applied to detect *Angiostrongylus cantonensis* DNA in cerebrospinal fluid (CSF)/clinical specimens [20–22].

Because parasitological/definitive diagnosis is rarely achieved, immunological and/or molecular diagnostic methods for the detection of *Angiostrongylus cantonensis* antibodies/antigens and/or nucleic acids have become widely accepted as the most appropriate diagnostic approach to support clinical diagnosis [2,12,23]. Nevertheless, the utilization of such tests is time-consuming, needs highly sophisticated, high-cost laboratory equipment and continuous electricity supply and is not suitable under clinical or field conditions in endemic regions [22].

The rapid global spread of *Angiostrongylus cantonensis* and the emerging occurrence of the infection have exposed the shortcomings of traditional/conventional diagnostics. This has spurred efforts to develop faster, simpler and more scalable platforms that can be decentralized for performing the on-site detection of *Angiostrongylus cantonensis* infection in order to allow prompt treatment decisions. At present, point-of-care immunoassays, such as the flow-through/lateral flow assay, are the best placed.

The LFA has recently attracted considerable interest because of its long shelf life and the fact that refrigeration is not required for storage. It is very well adapted for application in harsh field environments, and in remote regions. As such, an LFA in the form of an immunochromatographic test device (*Ac*AgQuick^{Dx}) based on the detection of a circulating *Angiostrongylus cantonensis*-derived antigen was established, using an anti-31 kDa *Angiostrongylus cantonensis* antibody line on the membrane strip as the capture reagent and anti-*Angiostrongylus cantonensis* polyclonal antibody conjugated to colloidal gold as the indicator reagent. The *Ac*AgQuick^{Dx} was initially evaluated for its diagnostic potential in this study.

2. Materials and Methods

2.1. Clinical Samples

A set of 20 individual CSF samples submitted to the Parasitology Laboratory, Department of Parasitology, Faculty of Medicine Siriraj Hospital, Bangkok, Thailand, for routine antibody testing (ELISA and/or immunoblot) of tissue-invading parasites, i.e., *Gnathostoma spinigerum*, *Angiostrongylus cantonensis* and *Taenia solium* metacestodes, were used for the present assessment. These CSF specimens were from clinically diagnosed cases with positive immunoblot tests for the presence of a 31-kDa band specific for *Angiostrongylus cantonensis* (n = 10; designated as CSF1–10) and clinically suspected cases with negative immunoblot tests for *Angiostrongylus cantonensis* infection (n = 5; CSF11–15), as well as CSF samples (representing other clinically related parasitic infections) with positive immunoblot tests showing a 24-kDa band specific for *Gnathostoma spinigerum* (n = 2; CSF16–17) and with ELISA-positive cases of *Taenia solium* neurocysticercosis (n = 3; CSF18–20). The CSF specimens were kept at −70 °C after the initial routine immunological investigations.

In addition, a total of 105 reference sera stock from the Parasitology Department of the Faculty of Medicine Siriraj Hospital, were also used for evaluation testing. Twenty-seven samples were from clinically diagnosed patients with detectable *Angiostrongylus cantonensis*-specific antibody in immunoblotting. The remaining 43 serum samples were from patients with other parasitic diseases, i.e., gnathostomiasis (n = 13), toxocariasis (n = 2), trichinellosis (n = 2), hookworm infection (n = 4), filariasis (n = 5), cysticercosis (n = 9), paragonimiasis (n = 2), opisthorchiasis (n = 3) and malaria (n = 3). These infections had been diagnosed using parasitological or serological methods. Additionally, 35 serum samples from normal healthy subjects (whose stool samples were without any intestinal parasitic infection) were included for testing.

All the 105 patient sera and 20 cerebrospinal fluids were collected from the leftover clinical samples stored separately at −70 °C. The same sets of archived CSF (n = 20) and serum (n = 97) samples had been tested with *Ac*DIGFA^{Ag} to detect the 31-kDa specific antigen of *Angiostrongylus cantonensis* [24]. Additionally, 8 serum specimens used in this study were from cases with positive anti-*Angiostrongylus cantonensis* antibodies detected using an immunoblot test. They were retrieved and re-tested using *Ac*AgQuick^{Dx} to confirm the presence of a 31-kDa *Angiostrongylus cantonensis* antigen. The use of stored leftover clinical CSF or serum samples for this study was approved by the Director of Siriraj Hospital, Faculty of Medicine Siriraj Hospital, Mahidol University.

2.2. Production and Purification of Polyclonal Antibodies

Procedures to produce rabbit immune sera against crude somatic extracts and purified 31-kDa glycoprotein of *Angiostrongylus cantonensis* were the methods previously described [24–27]. The rabbit antisera used in this study were those previously produced and stored in small aliquots at −70 °C. Anti-*Angiostrongylus cantonensis* and anti-31 kDa *Angiostrongylus cantonensis* immune sera collected 2 weeks after the last immunizing dose were used to establish the present *Ac*AgQuick^{Dx} device. From the previous immunoblot analysis, the crude *Angiostrongylus cantonensis* extracts were recognized using anti-*Angiostrongylus cantonensis* polyclonal antibody as multiple protein bands, including the 31-kDa antigenic band, whereas the anti-31 kDa antibody recognized a broad band with an approximate molecular mass of 31 kDa. The antisera were retrieved and purified using the Melon IgG Spin Purification Kit (Thermo Scientific, Waltham, MA, USA) according to the manufacturer's instructions. After purification, the viability of the purified rabbit antibodies was confirmed to ensure the test performance. The purified anti-31 kDa *Angiostrongylus cantonensis* IgG was used as an antigen-capture antibody, while the purified anti-*Angiostrongylus cantonensis* IgG was used for colloidal gold-labelling as a detection agent.

2.3. Preparation of the Lateral Flow Test Device

The colloidal gold-labelled polyclonal antibody against *Angiostrongylus cantonensis* was prepared and the immunochromatographic test device (designated *Ac*AgQuick^{Dx}) was then assembled as per the standard method by Serve Science Co., Ltd., Bangkok, Thailand. The lateral flow strip consisted of 4 components, i.e., the sample application pad, conjugate pad, antibody-immobilized nitrocellulose membrane and absorbent pad. In the detection zone/membrane, the purified anti-31 kDa *Angiostrongylus cantonensis* polyclonal antibody (1 mg/mL) was micro-sprayed (in a 1 mm wide line) at a flow rate of 0.1 µL per mm at the test line (T), and the goat anti-rabbit immunoglobublin-G (0.4 mg/mL) was sprayed at 0.1 μL per mm at the control line (C). The anti-*Angiostrongylus cantonensis* polyclonal antibody-coated colloidal gold probe was added to the conjugate pad as a detector reagent. The antibody-immobilized membrane and conjugate pad were dried overnight at 37 °C. The sample application pad, conjugate pad and absorbent pad were assembled with the antibody-immobilized nitrocellulose membrane on a laminated card by superposition/overlapping of the different pads, and was cut into strips (0.5 cm in width). Each test strip was housed in a protective plastic cassette, with a hole for sample application and a slot/window to display test results for interpretation, and then stored in a desiccated sealed aluminum foil-package at room temperature until used. The complete set of the *Ac*AgQuick^{Dx} kit consisted of an immunochromatographic cassette and chromatographic/chasing buffer.

2.4. Procedure of Lateral Flow Test Device

The test procedure was performed at room temperature. In the testing process, 25 μL of the test CSF/serum was applied slowly into the sample hole. After being completely absorbed, one drop (approximately 50 μL) of chromatographic/chasing buffer was added and allowed to be absorbed through the membrane to wash the excess colloidal gold-conjugated antibody/IgG. The result was interpreted within 15 min. The appearance of a red-colored band at the test (T) line and a red-colored band at the control (C) line indicated a positive result, whereas the absence of a red-colored band at the test (T) line and appearance of a red-colored band at the control (C) line indicated a negative result. The test was invalid when no red-colored bands appeared at either the T and C lines, or only one red-colored band appeared at the T line. All the parasite-infected patient CSF/serum samples and healthy control sera were tested twice with the *Ac*AgQuick^{Dx} device to confirm the reproducibility of the results.

3. Results

In this work, of the 10 CSF samples from patients showing clinical criteria of eosinophilic meningitis and positive results for a 31-kDa *Angiostrongylus cantonensis*-specific immunoblot band, three (CSF2, 3 and 7) had a positive *Ac*AgQuick^{Dx} reaction, showing positive red bands (with different color intensities) at the test (T) region within 15 min. Likewise, two (CSF12 and CSF15) of the five CSF samples from cases with clinical features of infection, but which were negative for *Angiostrongylus cantonensis*-specific antibodies, also displayed a positive *Ac*AgQuick^{Dx} test (Figure 1). Within the set of 27 patient sera with serologically confirmed *Angiostrongylus cantonensis* infection, four (Ac-7, Ac-15, Ac-22 and Ac-25) showed a positive reaction via *Ac*AgQuick^{Dx}, with visible pink bands at the T region (Figure 2). No positive *Ac*AgQuick^{Dx} reaction was observed in the other 23 sera with angiostrongyliasis. In cross-reactivity testing, all the 5 CSF samples and the 78 sera from gnathostomiasis (CSF = 2; serum = 13), toxocariasis (serum = 2), filariasis (serum = 5), trichinellosis (serum = 2), hookworm infection (serum = 4), cysticercosis (CSF = 3; serum = 9), paragonimiasis (serum = 2), opisthorchiasis (serum = 3), and malaria (serum = 3), as well as the 35 normal control sera from parasite-free individuals, were all negative for *Ac*AgQuick^{Dx}.

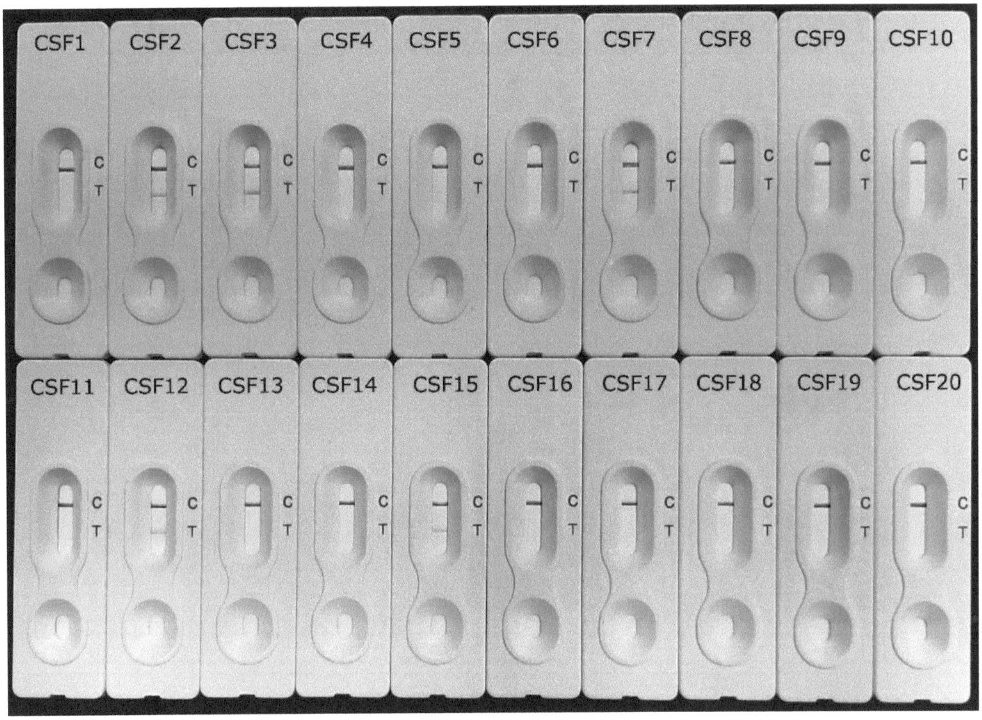

Figure 1. The *Ac*AgQuick^{Dx} detection results of 20 CSF samples from clinically diagnosed cases with positive immunoblot (antibody) test for angiostrongyliasis (CSF1–10), clinically suspected cases with negative immunoblot test (CSF11–15), and CSF samples with positive immunoblot (antibody test) for gnathostomiasis (CSF16–17), and with ELISA-positive cases of *Taenia solium* neurocysticercosis (CSF18–20). The appearance of a red band at the T line and a red band at the C line indicates a positive sample for the detection of the specific 31-kDa *Angiostrongylus cantonensis* antigen (CSF2, 3, 7, 12 and 15), whereas the absence of a red band at the T line and appearance of a red band at the C line indicates a negative sample (CSF1, 4, 5, 6, 8–11 and 13–14 and 16–20).

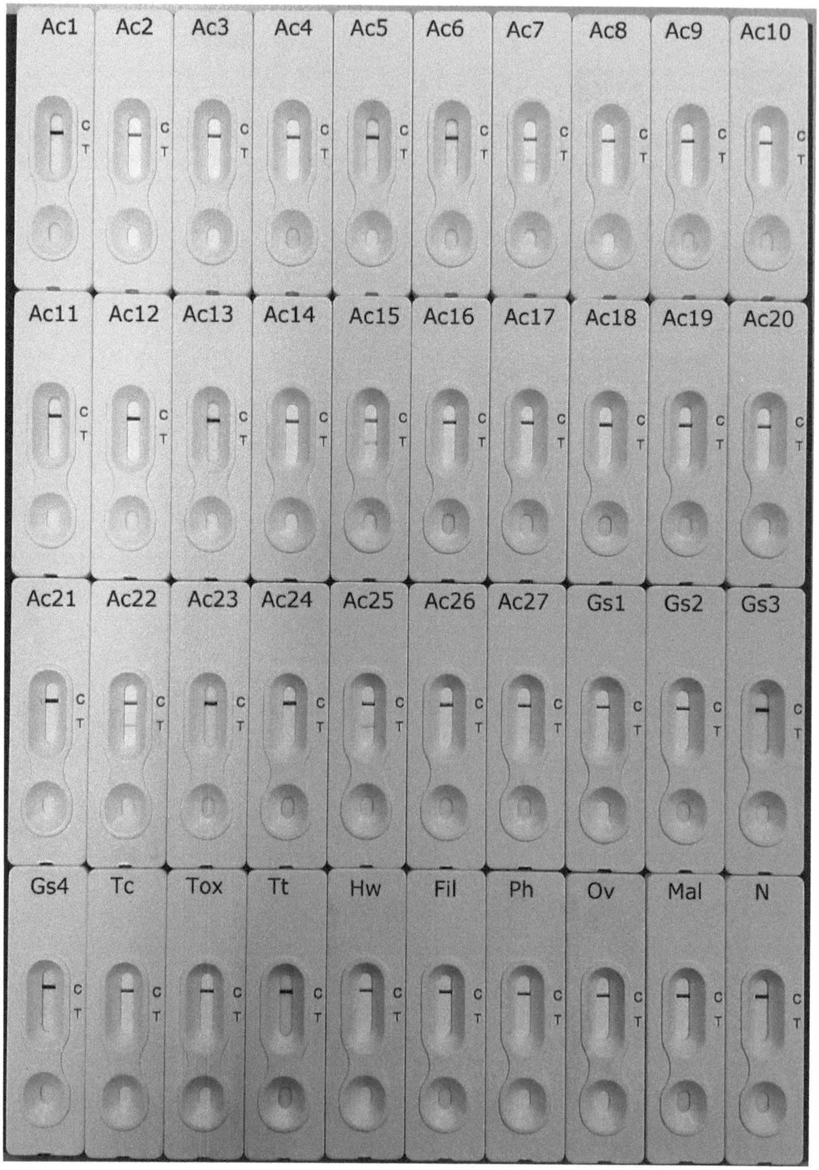

Figure 2. The *Ac*AgQuick^{Dx} test results of 27 serum samples from clinically diagnosed cases of angiostrongyliasis with positive immunoblot (antibody) test (Ac1–27), and the representative images of *Ac*AgQuick^{Dx} test evaluated using human sera with other heterologous parasitic infections (n = 43). Gs, gnathostomiasis; Tc, *Taenia solium* neurocysticercosis; Tox, toxocariasis; Tt, trichinellosis; Hw, hookworm infection; Fil, filariasis; Ph, paragonimiasis; Ov, opisthorchiasis; Mal, malaria; and N, normal healthy control (n = 35). The appearance of a red band at the T line and a red band at the C line indicates a positive result for detection of the specific 31-kDa antigen of *Angiostrongylus cantonensis* (Ac7, Ac15, Ac22 and Ac25). The absence of a red band at the T line and appearance of a red band at the C line indicates a negative test result (Ac1–6, Ac8–14, Ac16–21, Ac23–24, Ac26–27, Gs1–4, Tc, Tox, Tt, Hw, Fil, Ph, Ov, Mal and N).

4. Discussion

In clinical practice, the delayed diagnosis of *Angiostrongylus* eosinophilic meningitis due to its atypical symptoms is an important problem in many hospitals because it can cause fatal outcomes. Despite the recent introduction of a powerful and sensitive molecular diagnostic tool (metagenomic next-generation sequencing) for the specific identification of *Angiostrongylus cantonensis* DNA sequences in clinical specimens [28,29], the diagnosis of active/acute neuroangiostrongyliasis remains a challenge in resource-limited settings because of high operating costs and the requirement for laboratory infrastructure [29]. Other rapid and cost-effective assays for the point-of-need diagnosis are still required to meet the demand of appropriate tests for use in resource-poor settings or remote endemic regions.

In the early stages of infection, diagnosis based on antibody-based methods may lack sensitivity, especially during acute illness because seroconversion may take several weeks [30]. Alternatively, the detection of antigens has the advantage of detecting the presence of active *Angiostrongylus cantonensis* infection and the level of the infective burden. To date, only a few diagnostic tests have been reported for detecting specific *Angiostrongylus cantonensis* antigen in clinical samples. A rapid lateral flow immunoassay (LFIA), based on two monoclonal antibodies (12D5C12, 21B7B11) for detecting the specific antigens of *Angiostrongylus cantonensis*, revealed very high specificity (100%) and high sensitivity (91.1%) [31]. More recently, a rapid, non-enzymatic, dot immunogold filtration assay (*Ac*DIGFA^{Ag}) based on flow-through immunoassay, using purified antibodies against specific 31-kDa *Angiostrongylus cantonensis* antigen to detect a corresponding (specific) *Angiostrongylus cantonensis* antigen in the cerebrospinal fluid and serum samples from angiostrongyliasis patients, showed a diagnostic specificity of 100% [24].

The specific 31-kDa glycoprotein antigen of *Angiostrongylus cantonensis* has been suggested to be a potential immunological marker of early infection since this 31-kDa antigen is a component of the *Angiostrongylus cantonensis* excretory/secretory products [32]. This antigen has shown considerable promise as an immunodiagnostic target in an assay protocol based on the flow through/vertical flow principle [24]. However, the additional gold detector reagent for the test assay still needs refrigeration; as such, it is not completely applicable in the harsh field environment. As the relatively high concentration of the circulating 31-kDa antigen of *Angiostrongylus cantonensis* facilitates the development of other test platforms, this specific antigen was selected for application in a more field-friendly lateral flow assay. In addition, the gold-based, lateral flow immunochromatography approach to detect a specific 31-kDa glycoprotein of *Angiostrongylus cantonensis* (*Ac*AgQuick^{Dx} test) presented here was easily adapted from the previous method of *Ac*DIGFA^{Ag}, used to detect the specific 31-kDa *Angiostrongylus cantonensis* antigen [24].

In this study, the evaluation of the diagnostic potential of the *Ac*AgQuick^{Dx}, using a set of clinical samples from the reference stock, revealed 5 out of the 15 angiostrongyliasis CSF samples and 4 out of the 27 angiostrongyliasis serum samples to be positive, whereas none of the patient's samples with other heterologous parasitic infections (n = 48) and normal controls (n = 35) gave positive reactions. The overall diagnostic sensitivity and specificity of *Ac*AgQuick^{Dx} in detecting the 31-kDa antigens of *Angiostrongylus cantonensis* were 21.43% and 100%, respectively. The low sensitivity/positivity of the *Ac*AgQuick^{Dx} test in this study could possibly relate to delays in the diagnosis of the suspected active angiostrongyliasis cases, as the sensitivity of the antigen test depends on the phase of the disease, and on the presence of circulating *Angiostrongylus cantonensis* specific antigens in the CSF and blood. In the chronic cases, the antibody levels are expected to be elevated due to the persistence of antigen stimulation. The formation of antigen–antibody complexes, by the circulating antigens and antibodies, may inhibit the detection of antigens in these clinical samples [33]. Additionally, in those patients with a low intensity of infection, the positive antigen levels may be very close to the cut-off value.

The same set of archived serum samples (n = 19) in the present study was tested earlier using *Ac*Quick^{Dx} (specific antibody detection) with 100% sensitivity and 98.72% specificity [34]. In this study, only two out of the nineteen samples showed positive antigen results based on the examination/detection of the specific *Angiostrongylus cantonensis* antigen on the consecutive angiostrongyliasis serum samples using *Ac*AgQuick^{Dx}. On the other hand, two of the five CSF samples from clinically suspected cases (criteria of eosinophilic meningitis) with a negative immunoblot antibody test for *Angiostrongylus cantonensis* infection had positive antigen results using the present *Ac*AgQuick^{Dx} test. To overcome the limitations related to the early or late seroconversion phase of available clinical samples, the simultaneous use of both the rapid tests, *Ac*Quick^{Dx} (antibody detection) and *Ac*AgQuick^{Dx} (antigen detection), should be applied for accurate detection of neuroangiostrongyliasis.

In comparison to our present antigen-based immunochromatographic test, *Ac*AgQuick^{Dx}, the earlier *Angiostrongylus cantonensis*-derived antigen detecting *Ac*DIGFA^{Ag} test, with the identical pair of *Angiostrongylus cantonensis*-specific antibodies, revealed a slightly better sensitivity for the consecutive CSF sample tested (CSF1–15; Ac1-Ac19). One of six positive CSF samples tested with *Ac*DIGFA^{Ag} (CSF5) was negative with *Ac*AgQuick^{Dx}, whereas the two positive patient sera with *Ac*DIGFA^{Ag} gave concordant positive results with *Ac*AgQuick^{Dx}. This is likely that flow-through is an immunoconcentration assay and thus allows the detection of less abundant antigens in the samples.

Theoretically, in the antigen detection tests, monoclonal antibodies seem to be the better diagnostic reagents as they recognize a single epitope with high specificity. The *Ac*AgQuick^{Dx} in the present study showed 100% diagnostic specificity when tested with CSF/serum samples from patients with other clinically related parasitic infections. This was comparable with the earlier-developed *Ac*DIGFA^{Ag}, which also demonstrated a specificity of 100% [24]. Nevertheless, additional clinical samples from cases with heterologous parasitic infections, including other infectious diseases and cancers that may cause eosinophil abnormality in CSF and peripheral blood, need to be performed for a more rigorous evaluation of the specificity.

Furthermore, only patients from a single geographical region (Thailand) were included in this study, limiting the certainty that these data could apply to other endemic countries/regions with different geographical strains of *Angiostrongylus cantonensis* [35]. Whether the *Ac*AgQuick^{Dx} will also perform equally well in other *Angiostrongylus cantonensis* endemic regions remains to be evaluated. In addition, the question of the *Ac*AgQuick^{Dx} test yielding qualitative or semi-quantitative results needs to be addressed. The future goals for improving our *Ac*AgQuick^{Dx} will be focused on identifying new signal amplification strategies as well as the quantitative system.

5. Conclusions

Overall, our *Ac*AgQuick^{Dx} test based on the detection of a circulating 31-kDa *Angiostrongylus cantonensis*-derived antigen is a fast, portable and easy-to-use test device that meets the needs of laboratory testing in a variety of healthcare settings. The test allows the rapid immunological diagnosis of *Angiostrongylus cantonensis* infection to enable immediate clinical management decisions to be made at or near the site of patient care (at the point of care settings). It can also be a good alternative for use in the initial screening of neuroagiostrongyliasis in large-scale investigations in the field, where sophisticated equipment is lacking.

6. Limitations of the Work

There was a limitation to the current study because the immunological testing was carried out retrospectively, using stored frozen CSF and sera. It is possible that the use of fresh clinical samples may increase the test sensitivity.

Author Contributions: Conceptualization, project administration, methodology, investigation, data curation, formal analysis and writing—original draft preparation, P.E.; Conceptualization, methodology, resources and project administration, A.T.; Resources, methodology and reviewing, D.W.; Methodology, investigation and data curation, S.B.; Conceptualization, funding acquisition, supervision, writing—review and editing, H.-S.Y. All authors have read and agreed to the published version of the manuscript.

Funding: This work was supported in part by grants from the University of Malaya (H-5620009 and H-50001-00-A000025) to H.-S.Y.

Institutional Review Board Statement: This study was approved by the Ethics Committee of the Faculty of Medicine Siriraj Hospital, Mahidol University (approval number: Si 369/2015). The study came under the umbrella of "Research Developments in Emerging and Re-Emerging Infectious Diseases", Faculty of Medicine Siriraj Hospital (program number: 815/2557_EC1).

Informed Consent Statement: This study used leftover clinical samples submitted to the Parasitology Laboratory, Siriraj Hospital (under the Department of Parasitology, Faculty of Medicine Siriraj Hospital, Mahidol University) for routine immunological testing. An Informed Consent Statement is not needed. The leftover samples were stored separately at −70 °C without the patients' names or HN numbers.

Data Availability Statement: All the data applicable to this investigation are presented in this manuscript.

Acknowledgments: The authors wish to thank the Director of the Siriraj Hospital, Faculty of Medicine Siriraj Hospital, Mahidol University, for permission to publish this research work. We also appreciate the assistance by Nuntiya Monkong and Suphaluck Wankhom for the retrieval of CSF/serum samples kept at the Department of Parasitology, Siriraj Hospital.

Conflicts of Interest: The authors declare no conflict of interest.

References

1. Cross, J.H.; Chen, E.R. Angiostrogyliasis. In *Food Borne Parasitic Zoonoses*; Murrel, K.D., Fried, B., Eds.; Springer: New York, NY, USA, 2007; pp. 263–290.
2. Eamsobhana, P. *The Rat Lungworm Angiostrongylus cantonensis: Parasitology, Genetics and Molecular Phylogeny*, Revised 2nd ed.; Aksorn Graphic and Design Publishing House: Bangkok, Thailand, 2015.
3. Wang, Q.P.; Lai, D.H.; Zhu, X.Q.; Chen, X.G.; Lun, Z.R. Human angiostrongyliasis. *Lancet Infect. Dis.* **2008**, *8*, 621–630. [CrossRef] [PubMed]
4. Wang, Q.P.; Wu, Z.D.; Wei, J.; Owen, R.L.; Lun, Z.R. Human *Angiostrongylus cantonensis*: An update. *Eur. J. Clin. Microbiol. Infect. Dis.* **2012**, *31*, 389–395. [CrossRef] [PubMed]
5. Barratt, J.; Chan, D.; Sandaradura, I.; Malik, R.; Spielman, D.; Lee, R.; Marriott, D.; Harkness, J.; Ellis, J.; Stark, D. *Angiostrongylus cantonensis*: A review of its distribution, molecular biology and clinical significance as a human pathogen. *Parasitology* **2016**, *143*, 1087–1118. [CrossRef]
6. Cowie, R.H.; Ansdell, V.; Dunavan, C.P.; Rollins, R.L. Neuroangiostrongyliasis: Global spread of an emerging tropical disease. *Am. J. Trop. Med. Hyg.* **2022**, *107*, 1166–1172. [CrossRef] [PubMed]
7. Bhaibulaya, M. Morphology and taxonomy of major *Angiostrongylus* species of Eastern Asia and Australia. In *Studies on Angiostrongyliasis in Eastern Asia and Australia*; Cross, J.H., Ed.; NAMRU-2-SP-44; U.S. Naval Medical Research Unit: Taipei, Taiwan, 1979; pp. 4–13.
8. Turck, H.C.; Fox, M.T.; Cowie, R.H. Paratenic hosts of *Angiostrongylus cantonensis* and their relation to human neuroangiostrongyliasis globally. *One Health* **2022**, *15*, 100426. [CrossRef] [PubMed]
9. Punyagupta, S. Angiostrongyliasis: Clinical features and human pathology. In *Studies on Angiostrongyliasis in Eastern Asia and Australia*; Cross, J.H., Ed.; NAMRU-2-SP-44; U.S. Naval Medical Research Unit: Taipei, Taiwan, 1979; pp. 138–150.
10. Chotmongkol, V.; Sawanyawisuth, K. Clinical manifestation and outcome of patients with severe eosinophilic meningoencephalitis presumably caused by *Angiostrongylus cantonensis*. *Southeast Asian J. Trop. Med. Public Health* **2002**, *33*, 231–234.
11. Graeff-Teixeira, C.; da Silva, A.C.A.; Yoshimura, K. Update on eosinophilic meningoencephalitis and its clinical relevance. *Clin. Microbiol. Rev.* **2009**, *22*, 322–348. [CrossRef]
12. Ansdell, V.; Kramer, K.J.; McMillan, J.K.; Gosnell, W.L.; Murphy, G.S.; Meyer, B.C.; Blalock, E.U.; Yates, J.; Lteif, L.; Smith, O.A.; et al. Guidelines for the diagnosis and treatment of neuroangiostrongyliasis: Updated recommendations. *Parasitology* **2021**, *148*, 227–233. [CrossRef]
13. Eamsobhana, P.; Tungtrongchitr, A.; Wanachiwanawin, D.; Yong, H.S.; Mak, J.W. Characterization of a 31-kDa specific antigen from *Parastrongylus cantonensis* (Nematoda: Metastrongylidae). *Int. Med. Res. J.* **1998**, *2*, 9–12.
14. Nuamtanong, S. The evaluation of the 29 and 31 kDa antigens in female *Angiostrongylus cantonensis* for serodiagnosis of human angiostrongyliasis. *Southeast Asian J. Trop. Med. Public Health* **1996**, *27*, 291–296.
15. Maleewong, W.; Sombatsawat, P.; Intapan, P.M.; Wongkham, C.; Chotmongkol, V. Immunoblot evaluation of the specificity of the 29-kDa antigen from young adult female worms *Angiostrongylus cantonensis* for immunodiagnosis of human angiostrongyliasis. *Asian Pac. J. Allergy Immunol.* **2001**, *19*, 267–273. [PubMed]
16. Li, H.; Chen, X.G.; Shen, H.X.; Chen, D.X.; Qiu, Y.R.; Hu, X.J. Value of antigen with molecular mass of 32000 in immunodiagnosis of *Angiostrongylus cantonensis*. *J. First. Mil. Med. Univ.* **2005**, *25*, 380–383. (In Chinese)

17. Eamsobhana, P. Angiostrongyliasis in Thailand: Epidemiology and laboratory investigations. *Hawaii J. Med. Public Health* **2013**, *72*, 28–32. [PubMed]
18. Eamsobhana, P. Eosinophilic meningitis caused by *Angiostrongylus cantonensis*—A neglected disease with escalating importance. *Trop. Biomed.* **2014**, *31*, 569–578. [PubMed]
19. Jarvi, S.I.; Eamsobhana, P.; Quarta, S.; Howe, K.; Jacquier, S.; Hanlon, A.; Snook, K.; McHugh, R.; Tman, Z.; Miyamura, J.; et al. Estimating human exposure to rat lungworm (*Angiostrongylus cantonensis*) on Hawai'i Island: A pilot study. *Am. J. Trop. Med. Hyg.* **2020**, *102*, 69–77, accepted for publication. [CrossRef]
20. Eamsobhana, P.; Wanachiwanawin, D.; Dechkum, N.; Parsartvit, A.; Yong, H.S. Molecular diagnosis of eosinophilic meningitis due to *Angiostrongylus cantonensis* (Nematoda: Metastrongyloidea) by polymerase chain reaction-DNA sequencing of cerebrospinal fluids of patients. *Mem. Inst. Oswaldo Cruz.* **2013**, *108*, 116–118. [CrossRef]
21. Qvarnstrom, Y.; Xayavong, M.; da Silva, A.C.; Park, S.Y.; Whelen, A.C.; Calimlim, P.S.; Sciulli, R.H.; Honda, S.A.; Higa, K.; Kitsutani, P.; et al. Real-time polymerase chain reaction detection of *Angiostrongylus cantonensis* DNA in cerebrospinal fluid from patients with eosinophilic meningitis. *Am. J. Trop. Med. Hyg.* **2016**, *94*, 176–181. [CrossRef]
22. Verweij, J.J.; Stensvold, C.R. Molecular testing for clinical diagnosis and epidemiological investigations of intestinal parasitic infections. *Clin. Microbiol. Rev.* **2014**, *27*, 371–418. [CrossRef]
23. Eamsobhana, P.; Yong, H.S. Immunological diagnosis of human angiostrongyliasis due to *Angiostrongylus cantonensis* (Nematoda: Angiostrongylidae). *Int. J. Infect. Dis.* **2009**, *13*, 425–431. [CrossRef]
24. Eamsobhana, P.; Tungtrongchitr, A.; Yong, H.S.; Prasartvit, A.; Wanachiwanawin, D.; Gan, X.X. Sandwich dot-immunogold filtration assay (DIGFA) for specific immunodiagnosis of active neuroangiostrongyliasis. *Parasitology* **2021**, *148*, 234–239. [CrossRef]
25. Eamsobhana, P. Immunological studies on the rat lung-worm *Angiostrongylus cantonensis* (Nematoda: Metastrongylidae). Ph.D. Thesis, University of Malaya, Kuala Lumpur, Malaysia, 1994.
26. Eamsobhana, P.; Yong, H.S.; Mak, J.W.; Wattanakulpanich, D. Detection of circulating antigens of *Parastrongylus cantonensis* in human sera by dot-blot ELISA and sandwich ELISA using monoclonal antibody. *Southeast Asian J. Trop. Med. Public Health* **1997**, *28*, 624–628.
27. Eamsobhana, P.; Yoolek, A.; Suvouttho, S.; Suvuttho, S. Purification of a specific immunodiagnostic *Parastrongylus cantonensis* antigen by electroelution from SDS-polyacrylamide gels. *Southeast Asian J. Trop. Med. Public Health* **2001**, *32*, 308–313.
28. Fan, X.W.; Wang, Q.; Li, P.Q.; Ai, B.; Song, Y.L.; Peng, Q.Y.; Wang, H.L. The diagnostic value of metagenomic next-generation sequencing in *Angiostrongylus cantonensis* encephalitis/meningitis. *J. Behav. Brain Sci.* **2021**, *11*, 216–226. [CrossRef]
29. Liu, J.; Tao, J.; Chen, W.; Wang, T.; Chen, X.; Shen, M.; Ou, Q.; Zhang, Y.; Ding, Y.; Wu, J.; et al. The application of metagenomic next-generation sequencing for *Angiostrongylus* eosinophilic meningitis in a pediatric patient: A case report. *Front. Public Health* **2022**, *10*, 1003013. [CrossRef] [PubMed]
30. Slom, T.J.; Cortese, M.M.; Gerber, S.I.; Jones, R.C.; Holtz, T.H.; Lopez, A.S.; Zambrano, C.H.; Sufit, R.L.; Sakolvaree, Y.; Chaicumpa, W.; et al. An outbreak of eosinophilic meningitis caused by *Angiostrongylus cantonensis* in travelers returning from the Caribbean. *N. Engl. J. Med.* **2002**, *346*, 668–675. [CrossRef] [PubMed]
31. Chen, M.X.; Chen, J.X.; Chen, S.H.; Huang, D.N.; Ai, L.; Zhang, R.L. Development of lateral flow immunoassay for antigen detection in human *Angiostrongylus cantonensis* infection. *Korean J. Parasitol.* **2016**, *54*, 375–380. [CrossRef] [PubMed]
32. Eamsobhana, P.; Dechkum, N. Recognition of a diagnostic antigen from excretory/secretory products of adult *Parastrongylus cantonensis* by infected human sera. *Trop. Biomed.* **2000**, *17*, 103–108.
33. Shih, H.H.; Chen, S.N. Immunodiagnosis of angiostrongyliasis with monoclonal antibodies recognizing a circulating antigen of mol. wt. 91,000 from *Angiostrongylus cantonensis*. *Int. J. Parsitol.* **1991**, *21*, 171–177. [CrossRef]
34. Eamsobhana, P.; Tungtrongchitr, A.; Wanachiwanawin, D.; Yong, H.S. Immunochromatographic test for rapid serological diagnosis of human angiostrongyliasis. *Int. J. Infect. Dis.* **2018**, *73*, 69–71. [CrossRef]
35. Cross, J.H. Experimental studies on *Angiostrongylus* species and strains in monkeys and laboratory animals. In *Studies on Angiostrongyliasis in Eastern Asia and Australia*; Cross, J.H., Ed.; NAMRU-2-SP-44; U.S. Naval Medical Research Unit: Taipei, Taiwan, 1979; pp. 118–137.

pathogens

Article

Comparison of Antibody Isotype Response to *Angiostrongylus cantonensis* in Experimentally Infected Rats (*Rattus norvegicus*) Using Hawai'i 31 kDa Antigen in an Indirect ELISA

Argon Steel, Lisa Kaluna, John Jacob and Susan Jarvi *

Department of Pharmaceutical Sciences, Daniel K. Inouye College of Pharmacy, University of Hawai'i at Hilo, 200 West Kawili St., Hilo, HI 96720, USA

* Correspondence: jarvi@hawaii.edu; Tel.: +1-(808)-932-7701

Abstract: Neuroangiostrongyliasis (NAS) is an emerging tropical disease in humans and some animals which is caused by infection with the parasitic nematode *Angiostrongylus cantonensis*. It is the leading cause of eosinophilic meningitis worldwide. Diagnoses in humans and susceptible animals are generally presumptive and easily confused with other central nervous system disorders. The 31 kDa antigen is currently the only NAS immunodiagnostic assay that has achieved 100% sensitivity. However, little is known about the humoral immune response against the 31 kDa antigen in NAS infections, which would be critical for widespread adoption of this assay. We used the Hawai'i 31 kDa isolate in an indirect ELISA assay to confirm the presence of immunoglobulin IgG, IgM, IgA, and IgE isotypes in six-week post-infection plasma from lab-reared rats infected with 50 live, third-stage, *A. cantonensis* larvae isolated from a wild *Parmarion martensi* semi-slug. Our results confirmed the presence of all four isotypes against the Hawaii 31 kDa isolate, with sensitivity ranging from 22–100%. The IgG isotype showed 100% sensitivity in detecting *A. cantonensis* infection, which validates the use of IgG indirect ELISA with 31 kDa antigen as an effective immunodiagnostic assay for rats six weeks post-infection. Given each isotype may be present at different times during NAS infections, our data provides preliminary information on the humoral immune response to *A. cantonensis* infection in lab-reared rats and serves as a baseline for future studies.

Keywords: *Angiostrongylus cantonensis*; angiostrongyliasis; rat lungworm; serological detection; ELISA; 31 kDa

check for updates

Citation: Steel, A.; Kaluna, L.; Jacob, J.; Jarvi, S. Comparison of Antibody Isotype Response to *Angiostrongylus cantonensis* in Experimentally Infected Rats (*Rattus norvegicus*) Using Hawai'i 31 kDa Antigen in an Indirect ELISA. *Pathogens* **2023**, *12*, 625. https://doi.org/10.3390/pathogens12040625

Academic Editor: Anastasia Diakou

Received: 13 December 2022
Revised: 13 April 2023
Accepted: 18 April 2023
Published: 21 April 2023

1. Introduction

The nematode *Angiostrongylus cantonensis* (rat lungworm) is a food and water-borne zoonotic parasite causing neuroangiostrongyliasis (NAS), an emerging tropical infectious disease and the leading cause of eosinophilic meningitis worldwide [1,2]. In the life cycle of *A. cantonensis*, rats are the obligate definitive host, with adult worms reproducing in the pulmonary arteries, and gastropods are the obligate intermediate host [3]. Humans and other mammals, such as horses and dogs, can be accidental hosts, in which *A. cantonensis* does not reproduce but does invade the central nervous system (CNS), causing painful and devastating symptoms [3–6]. Symptoms of NAS are frequently mistaken for other central nervous system disorders, leading to frequent misdiagnoses in humans and animals [7,8].

A reliable laboratory diagnostic assay that can be used in conjunction with patient history and clinical symptoms has been a longstanding goal for NAS. The gold standard for diagnosis of *A. cantonensis* infection is the microscopic identification of whole larvae in the cerebrospinal fluid (CSF) [9]. With advances in technology, the Centers for Disease Control and Prevention (CDC) currently tests human CSF for *A. cantonensis* DNA using a real-time polymerase chain reaction (PCR) assay [10], and a few laboratories offer real-time PCR testing for animals. The most recent real-time PCR assay for NAS [11] meets important laboratory diagnostic criteria, such as having high sensitivity, and will likely be the assay of choice in the near future.

Nevertheless, while no laboratory diagnostic is perfect in all circumstances, numerous problems plague both the identification of whole larvae in CSF and real-time PCR, which complicate testing and delay or prevent patient diagnosis. For example, collection of CSF can be difficult to obtain, as some patients, doctors, or veterinarians are hesitant to perform a lumbar puncture. Both microscopic identification and real-time PCR have a high failure rate due to the scarcity of their targets and often require repeated lumbar punctures for positive identification [9]. As a result, most diagnoses are presumptively based on compatible history, characteristic physical findings, and evidence of high eosinophil counts [7,12]; thus, ongoing research is searching for a less invasive and more robust laboratory diagnostic. As an immunodiagnostic test, the measurement of antibody titers has the potential for high sensitivity while also overcoming some barriers given that blood collection is less invasive, antibodies can reach levels easily detected, and the advantage that previous exposure to the pathogen can be retrospectively detected.

For decades, a variety of antibody detection methods have been investigated for NAS, with most research focused on the enzyme-linked immunosorbent assay (ELISA). For example, Cross [13] developed an ELISA assay using *A. cantonensis* crude antigen isolated from fourth-stage larvae recovered from rat brains. When tested against other helminths, the ELISA values for *A. cantonensis* were higher, yet cross-reactivity was still significant. Subsequent research has been directed toward finding discrete antigens with greater specificity through various methods of purification of *A. cantonensis* proteins [14–16]. A number of these studies found that serum antibodies from human patients with NAS specifically recognized 29 kDa and 31 kDa proteins present in crude antigen of adult worms, with further research showing greater specificity for the 31 kDa protein [6,12]. After purifying the 31 kDa antigen using electrophoresis in a 12% SDS-polyacrylamide gel, Eamsobhana et al. [17] were able to achieve 100% sensitivity and specificity when testing serum from human patients with active *A. cantonensis* infections. Some research suggests that native antigens isolated from geographically local *A. cantonensis* may enhance the detection of exposure to the parasite or prior NAS infections, particularly in non-endemic areas [10,18]. Despite having been validated over a decade ago, there is a lack of basic immunological information about the humoral immune response to the 31 kDa antigen that is pertinent for a robust laboratory diagnostic. Knowing how antibody isotype titers change over time, in different tissues, following various infection loads, or in different hosts would likely be helpful to distinguish either early diagnosis or between active and prior infections.

Investigations into the NAS humoral immune response are most easily conducted using a laboratory-reared animal model to eliminate or reduce cross-reactivity from prior infections, as well as to gain the ability to control the dose and timing of infection. While the definitive host (the rat) may have different humoral immune responses to NAS than accidental hosts (humans or other mammals), the rat is a valuable and widely used model for investigating humoral immune response and assay validation. Likewise, while the IgG antibody is the most widely used isotype for ELISA assays in both NAS and non-NAS infections, other isotypes may be informative as a laboratory diagnostic, particularly early in a NAS infection when IgG levels are lower [19–22]. IgM is the first to appear following initial exposure to an antigen and might be particularly informative for early NAS diagnosis. IgA plays a role in mucous membranes and could be activated early in an infection when *A. cantonensis* is in the gastrointestinal tract. As a critical part of the immune defense against parasitic worms, IgE may also be informative. In fact, immunoglobins G, M, A, and E are all detectable in the sera and CSF of humans with acute, natural NAS infections [16,23]. In this study, we optimized methods developed for human sera [24] to confirm the presence of IgG, IgM, IgE, and IgA isotypes in six-week post infection (PI) rat plasma against the Hawaii 31 kDa isolate using an indirect ELISA assay.

2. Materials and Methods

2.1. Rat Plasma

The plasma for this study was collected from rats used to validate the use of propidium iodide stain as an in vitro death assay [25]. There were several advantages of using this plasma source. First, laboratory strains of rats were used in the study instead of wild rats, minimizing the likelihood of cross-reactivity or acquired immunity resulting from prior infections. Additionally, appropriate experimental groups were selected for the ELISA study, enabling the inclusion of both positive and negative controls, a suitable number of replicates within experimental groups, and sufficient plasma for testing. While plasma collection was limited to the six-week PI necropsy, this time point is a good starting point for ELISA validation and testing in rats, as IgG titers are expected to be high. This is because *A. cantonensis* completes its life cycle by this time, and IgG against crude antigen is typically present between 10–30 days and peaks by 50–60 days PI [19–22].

Briefly, 40 7-month-old Wistar IGC outbred laboratory strains of *Rattus norvegicus* were obtained from Charles River Labs (Raleigh, NC, USA) and maintained as described at the USDA-APHIS Wildlife Services National Wildlife Research Center (NWRC) Hawai'i Field Station, Hilo, HI, USA. Rats were gavaged under sedation as follows: Uninfected control rats were fed 1 mL of dH_2O only, without larvae, while other rats were gavaged with 1 mL of dH_2O containing 50 *A. cantonensis* L3, either live or dead, stained with propidium iodide or unstained. Thus, experimental groups consisted of: (1) live-stained larvae (3 male/2 female rats); (2) live-unstained larvae (2 male/2 female rats; one female was found dead in the cage, 40 days PI and before euthanasia and cardiac bleed); (3) killed-stained larvae (5 male/5 female rats); (4) killed-unstained larvae (5 male/5 female rats); and (5) uninfected controls (5 male/5 female rats) for a total of 39 rats. All *A. cantonensis* larvae were obtained from a single semi-slug (*Parmarion martensi*) collected from the University of Hawai'i at Hilo campus.

Rats were humanely euthanized six weeks PI, and necropsies were conducted as described [25]. Briefly, following euthanasia, whole blood samples (up to 2 mL) were collected via cardiac puncture of the right ventricle using a tuberculin syringe and transferred to heparinized collection vials for plasma isolation, transferred to sterile screw-capped tubes, and stored at −80 °C until testing with ELISA. In addition, selected organs were dissected: heart and lung were examined for adult *A. cantonensis*, the brain for the presence of larvae, and the lungs were examined for evidence of granulation (see Supplementary Materials Data Table S1).

2.2. Hawai'i 31 kDa Isolation

The Hawai'i 31 kDa isolate prepared by Jarvi et al. [26] with methods based on Eamsobhana et al. [17,27] was used for ELISA testing in this study. Briefly, adult female worms were harvested from the heart and lung tissue of infected rats collected from East Hawai'i Island between January through February 2017, washed with 1X PBS buffer, and stored at −80 °C in 1X protease inhibitor (Biochem Cocktail set V EDTA-Free, Thermo Scientific, Waltham, MA, USA) in 0.01 M PBS (Life Technologies, Grand Island, NY, USA). Frozen worms were homogenized manually in 1X protease inhibitor diluted in 1X PBS with a glass homogenizer, then sonicated (QSonica) on ice between 3 to 10 times in 3 s intervals with a 20 s rest period between each cycle to prevent overheating. The homogenized sample was stored at 4 °C overnight, centrifuged, and collected as soluble antigen (supernatant). The recovered antigen was quantified using a Coomassie Plus (Bradford) Assay Kit (ThermoFisher Scientific, Waltham, MA USA), then separated on a series of 12% SDS-polyacrylamide gels using electroelution to isolate the 31 kDa targeted antigen. Sections of gel containing the 31 kDa proteins were identified using a Hi-Mark^{TM} Pre-stained Protein Standard ladder, and these gel sections were excised manually. The gel slice was minced and then eluted with a Model 422 Electro-Eluter (Bio-Rad Laboratories, Hercules, CA, USA). Electroeluated protein was desalted and concentrated by ultrafiltration using Amicon Ultra-2 Centrifugal Filter Devices (MilliPore Sigma, Burlington, MA, USA) and quantitated using High

Sensitivity Protein 250 chips in an Agilent 2100 Bioanalyzer System (Agilent Technologies, Santa Clara, CA, USA) as having a protein concentration of 1.69 µg/µL. Quantified proteins were pooled and stored in 1.5 mL low protein binding microcentrifuge tubes (Eppendorf, Hauppauge, NY, USA) at −80 °C.

2.3. Indirect ELISA

ELISA was based on the methods described in Jarvi et al. [26], with some modifications. Antigen, plasma, and conjugate concentrations were determined using checkerboard titration. Plasma from rats fed live-unstained larvae was used as positive controls, and plasma from uninfected rats was used as negative controls; infection status was confirmed for all experimental groups through necropsy. Primary antibodies were derived via titration of positive and negative plasma controls with a range of 1:50 to 1:1600, resulting in final concentrations of 1:100 for IgG and IgM and 1:50 for IgA and IgE. Concentrations of secondary antibody (all conjugated with horseradish peroxidase) were derived from titration from an original range of 1:450 to 1:100,000, with final concentrations of 1:5000 for IgG-Fc Fragment (Bethyl Labs, A110-136P, Lot 30) and IgM (Bethyl Labs, A110-100P, Lot: 46) and 1:1250 for IgA (Bethyl Labs, A110-102P, Lot: 40) and IgE (Invitrogen, No. SA5-10256, Lot XE3581801). Antigen concentrations of 0.25 µg/well for IgG and IgM and 0.5 µg/well for IgA and IgE and were derived from a titration of 1.0 to 0.0313 µg/well, quantitated from 31 kDa antigen with a concentration of 1.69 µg/µL (see above).

Each plate included at least two positive controls, two negative controls, and a single carbonate buffer control or blank (without antigen). All samples and controls were run in triplicate. Flat bottom 96-well Immulon 4HBX microtiter plates (ThermoScientific) were initially coated with 31 kDa antigen derived from *A. cantonensis* and diluted with 0.05 M BupH carbonate–bicarbonate buffer (ThermoScientific, #28382) at pH 9.4 and refrigerated overnight at 4 °C. All washing steps were conducted on a 405 Select TS microplate washer (BioTek, Winooski, VT, USA). After coating, plates were washed four times with 300 µL PBS–0.05% Tween 20 (PBS-T) (pH 7.4) with a 2 min pause after every other wash. Plates were then blocked (125 µL/well) with 5% BLOTTO (nonfat dry milk powder) in PBS-Tween (PBS-T) for 2 h at room temperature with gentle rocking. The blocking solution was removed with no washing, then plasma samples diluted with 2.5% BLOTTO in PBS-T were added to appropriate wells (100 µL/well) and incubated for 2 h at 37 °C with gentle shaking. After incubation with primary antibody, plates were washed six times with 300 µL PBS-T, with a 2 min pause after every other wash. Horseradish peroxidase (HRP)-conjugated goat anti-rat secondary antibody diluted in 2.5% BLOTTO in PBS-T was added (100 µL/well), and plates were incubated for 1 h at 37 °C with gentle shaking. Plates were then washed six times with PBS-T, pausing every other wash. TMB-solubilized substrate solution (TMB One®; Promega, Madison, WI, USA) for HRP was added (100 µL/well). Absorbance was monitored on an µQuant microplate reader (BioTek) at 650 nm and is reported as optical density (OD). Upon the positive controls reaching a maximum OD of ≤0.6, the reaction was stopped with 1 N HCl (100 µL/well), allowed to equilibrate for 5 min with gentle rocking, then OD read at 450 nm.

2.4. Indirect ELISA Data Analysis

The cutoff value to distinguish positive vs. negative infection status was determined for each immunoglobulin class by taking the mean +3 standard deviations $\left(X \pm 3SD\right)$ of the OD values of the negative control plasma samples [17]. Thus, a sample was considered positive if the OD value exceeded the cutoff value. Sensitivity, defined as the percentage of individuals with a given condition whom the assay identifies as positive for that condition [28], was calculated as follows,

$$Sensitivity = \frac{N_{tp}}{\left(N_{tp} + N_{fn}\right)}$$

where N_{tp} and N_{fn} denote the number of true positive results and the number of false-negative results, respectively. In this paper, we define true positives as plasma samples collected from rats confirmed as having *A. cantonensis* infection through necropsy. False negatives are defined as plasma samples confirmed as having *A. cantonensis* infection through necropsy but whose OD value fell below the positive cutoff. True negatives are defined as plasma from uninfected rats. Spearman's tests were used to measure correlations of absorbance (OD levels) with rat sex and weight, the staining of larvae fed to rats, and the number of adult worms found during necropsy, for each antibody isotype. All analyses were performed using Minitab Statistical Software v21.3.

3. Results

For each isotype, optical density (OD) results showed that neither propidium iodide staining nor rat sex status produced significant differences within the groups of rats gavaged with either live or killed larvae (Table 1 and Figures 1 and 2); thus, the results are discussed as simply live vs. killed.

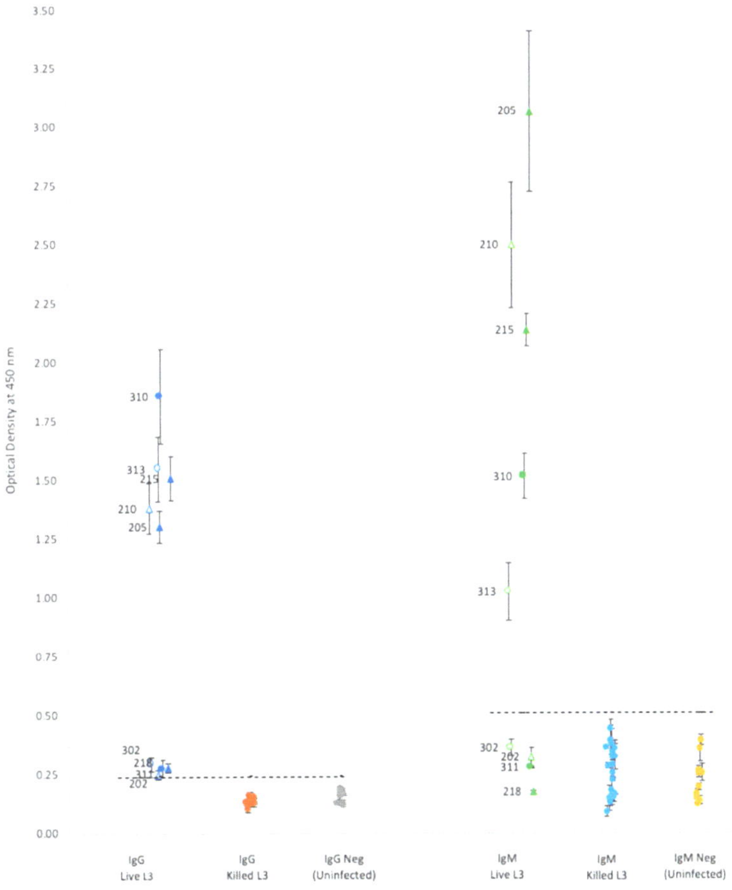

Figure 1. A comparison of optical density (*Y*-axis) by experimental groups (*X*-axis) using IgG and IgM conjugate, respectively. IgG and IgM are comparable due to their common assay concentrations; primary antibody concentration (1:1000), secondary antibody concentration (1:5000), and antigen concentration (0.25 µg/well). Individual rat ID numbers are as indicated. Dashed lines indicate positive cutoff. ▲ = Male rats fed stained live L3, △ = Male rats fed unstained live L3, ● = Female rats fed stained live L3, ○ = Female rats fed unstained live L3.

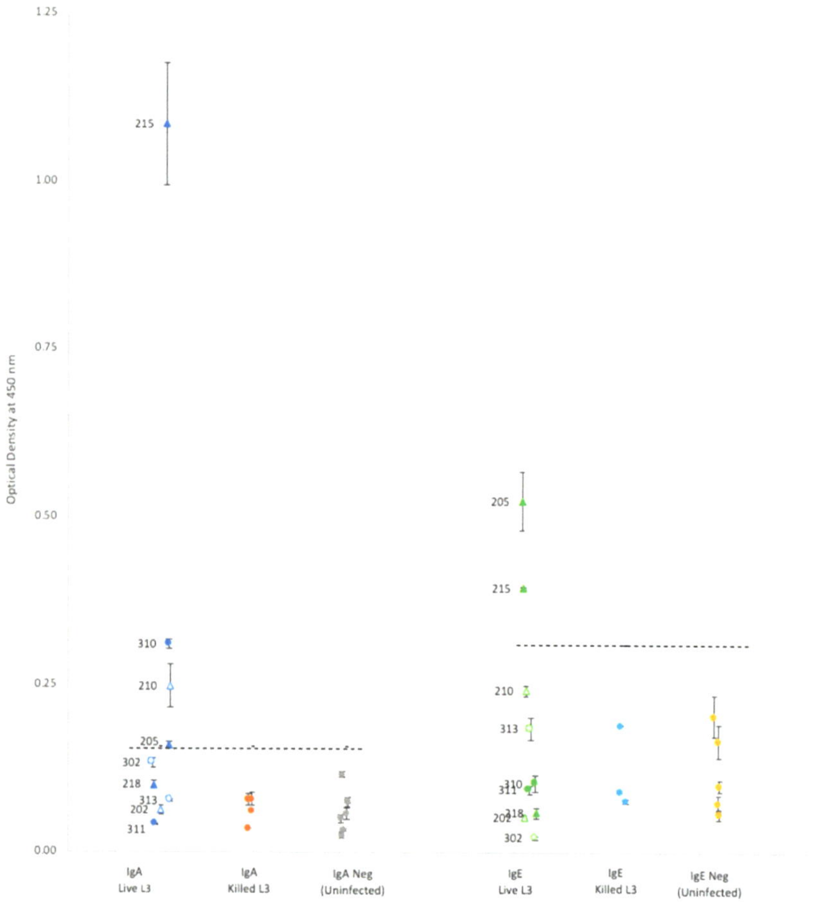

Figure 2. A comparison of optical density (*Y*-axis) by experimental groups (*X*-axis) using IgA and IgE conjugate, respectively. IgA and IgE are comparable due to their common assay concentrations; primary antibody concentration (1:50), secondary antibody concentration (1:1250), and antigen concentration (0.5 µg/well). Individual rat ID numbers are as indicated. Dashed lines indicate positive cutoff. ▲ = Male rats fed stained live L3, △ = Male rats fed unstained live L3, • = Female rats fed stained live L3, ○ = Female rats fed unstained live L3.

Table 1. Pairwise Spearman Correlation (*r*).

	Mean OD: IgG	Mean OD: IgM	Mean OD: IgA	Mean OD: IgE
Stain	−0.066	0.045	0.206	0.193
Sex (all rats)	0.048	−0.107	−0.407	−0.265
Proportional weight diff.	−0.317	−0.017	0.033	0
Total worms in rat lungs	−0.038	−0.5	−0.517	−0.167

Plasma from rats gavaged with live larvae (9 rats) generally showed markedly higher OD values than those from rats gavaged with killed larvae (20 rats) or the uninfected control group (10 rats). However, using the mean of the control group +3 standard deviations as a cutoff value for determining positive vs. negative infection status provided different results for each antibody. Positive OD cutoff values for IgG, IgM, IgA, and IgE were 0.238, 0.505,

0.156, and 0.309, respectively (dashed lines in Figures 1 and 2). This positive cutoff value resulted in a sensitivity for each isotype of 100%, 56%, 44%, and 22%, respectively. The mean OD values for rats fed killed larvae all fell below the positive cutoff in each isotype, indicating their negative infection status.

Each isotype showed a wide range of OD readings in the plasma of rats fed live larvae; however, no significant correlation (Spearman's test) was detected between ELISA OD levels and the rat's proportional weight difference, or the number of adult worms found in the rat's lungs upon autopsy regardless of antibody isotype (Table 1).

4. Discussion

The results of this study confirm the presence of IgG, IgM, IgE, and IgA isotypes against the Hawai'i 31 kDa isolate in at least some six-week PI rat plasma, with wide variation in the sensitivity of each isotype. The 22–100% range in sensitivity observed in this study differs from human studies conducted with crude antigen, which found each isotype had a roughly 70–98% sensitivity during acute infection [16]. Additional research is needed to determine the cause of this discrepancy. Similarities in the humoral immune response exist across different antigens, hosts, parasite burdens, diagnostic methods, and even infections with different *Angiostrongylus* species. For example, studies using a variety of diagnostic methods, parasite burdens, and stages of *A. cantonensis* for crude antigen preparation have shown that during primary infection, total antibodies and the IgG isotype against crude *A. cantonensis* antigens can be detected in rat sera as early as 10–30 days PI and peak at 50–60 days PI [19–22]. Our study found 100% sensitivity of IgG against the Hawai'i 31 kDa isolate at 42 days PI, which correlates with this detection window. Similarly, serum antibodies in canine *A. vasorum* infections are also detectable at 27 days PI and peak at 55–57 days PI and remain detectable at 83–84 days PI [29]. The consistency of antibody detection across different *A. cantonensis* antigens, different *Angiostrongylus* species, and in different hosts, indicates that IgG against the Hawai'i 31 kDa isolate may be a good biomarker for NAS infections of accidental hosts as early as 10 days PI through at least 60 days PI, possibly longer. Additional research is needed to determine the sensitivity of the IgG-Hawai'i 31 kDa ELISA assay throughout this potential testing window.

Given that early treatment may result in better patient outcomes, possibly avoiding chronic sequelae, a laboratory diagnostic that can detect NAS acute infections within the symptom onset window could result in better patient outcomes [9]. Presentation of NAS symptoms may occur before the potential IgG testing window described above, with humans exhibiting symptoms between 7 and 21 days PI [9] and canines showing symptoms at approximately 11 days PI [30]. While NAS infections have been confirmed in equines, the time of symptom onset is unknown [30]. Although Takai et al. [20] found both IgG and IgM antibodies in rats at 10 days PI, this early detection has not been confirmed, with Kanbara et al. failing to detect an IgG antibody response in rats until 20–30 days PI [21]. Given the IgM isotype is the first to appear in a primary infection, and our results showed relatively high OD readings in 56% of rats gavaged with live larvae, the IgM isotype against the Hawai'i 31 kDa isolate could be an informative laboratory diagnostic during NAS symptom onset period. Unfortunately, it seems the change in IgA or IgE titers over time in NAS infections has not been investigated. While our study was able to detect IgA and IgE in some rats gavaged with live larvae (44% and 22%, respectively), additional research is needed to determine if the IgA or IgE isotypes against the Hawai'i 31 kDa isolate could be informative for NAS infections.

Our results showed distinctive bimodal clustering in IgG and IgM OD values (Figure 1) for rats with active infections, but the reasons for this are unclear. Although two studies have found that the total antibody concentration, IgG and IgM, in rat serum varies according to larval load at time of infection [20,21], the positive plasma used in this study came from rats that were all gavaged with 50 live larvae. Our results differ from ELISA results of active human NAS infections, which show a relatively uniform distribution pattern in OD values [16]. Moreover, no correlation was found between titers and the number of adult *A.*

cantonensis found in rat pulmonary arteries. Additional research is needed to determine what causes the variation in isotype OD values in the Hawai'i 31 kDa ELISA assay.

In our study, there was no detectable antibody response in rats fed killed larvae. However, this might be a consequence of the methanol used to kill the larvae, which may have denatured the 31 kDa proteins on the surface of the larvae. Thus, additional research is needed to determine under what conditions dead larvae can generate a humoral immune response.

This study is the first step toward a more in-depth understanding of the humoral immune response against the *A. cantonensis* 31 kDa antigen. As seen with *A. vasorum* [29], it might be possible that blood antibody titers using ELISA offers a wider window than other high-sensitivity laboratory diagnostic methods such as real-time PCR of blood, and thus warrant further investigation.

Supplementary Materials: The following supporting information can be downloaded at: https://www.mdpi.com/article/10.3390/pathogens12040625/s1, Table S1: Supplementary Data-DAAutopsy vs. OD.xlsx.

Author Contributions: Conceptualization, A.S. and S.J.; Methodology, A.S., L.K. and S.J.; Investigation, A.S.; Writing—Original Draft Preparation, A.S.; Writing—Review and Editing, A.S., L.K., J.J. and S.J.; Supervision, S.J.; Funding Acquisition, S.J. All authors have read and agreed to the published version of the manuscript.

Funding: This work was supported by the Hawai'i State Legislature; by the Daniel K. Inouye College of Pharmacy, University of Hawai'i at Hilo, Hilo, HI.

Institutional Review Board Statement: All animal procedures were conducted according to the Guidelines of the American Society of Mammologists for the use of mammals in research (Sikes et al., 2011) and following approved Institutional Animal Care and Use Committee protocols (USDA NWRC QA-2879) and (University of Hawai'i) exempt TEX 18-007. This study was approved by the University of Hawai'i Institutional Biosafety Committee.

Informed Consent Statement: Not applicable.

Data Availability Statement: All data supporting the reported results are provided in the manuscript.

Acknowledgments: We wish to thank John Berestecky, Kapiolani Community College, for his many practical suggestions on ELISA, particularly checkerboard titrations, Marcos Gorresen, U.S. Geological Survey, Pacific Island Ecosystems Research Center, for his advice on aspects of statistical analysis, and Yuwei Wang for her assistance in running ELISAs. We also thank USDA-APHIS Wildlife Services, National Wildlife Research Center, Hawaii Field Station, Hilo HI USA, for research assistance and vivarium support in completing the initial study [25] in which the plasma used in this study was collected.

Conflicts of Interest: The authors declare there are no conflict of interest. The funders had no role in the design of the study; in the collection, analyses, or interpretation of data; in the writing of the manuscript; or in the decision to publish the results.

References

1. Graeff-Teixeira, C.; da Silva, A.C.A.; Yoshimura, K. Update on Eosinophilic Meningoencephalitis and Its Clinical Relevance. *Clin. Microbiol. Rev.* **2009**, *22*, 322–348. [CrossRef] [PubMed]
2. Prociv, P.; Turner, M. Neuroangiostrongyliasis: The "Subarachnoid Phase" and Its Implications for Anthelminthic Therapy. *Am. J. Trop. Med. Hyg.* **2018**, *98*, 353–359. [CrossRef] [PubMed]
3. Alicata, J.E. Biology and Distribution of the Rat Lungworm, *Angiostrongylus cantonensis*, and Its Relationship to Eosinophilic Meningoencephalitis and Other Neurological Disorders of Man and Animals. *Adv. Parasitol.* **1965**, *3*, 223–248. [CrossRef] [PubMed]
4. Liu, E.W.; Schwartz, B.S.; Hysmith, N.D.; DeVincenzo, J.P.; Larson, D.T.; Maves, R.C.; Palazzi, D.L.; Meyer, C.; Custodio, H.T.; Braza, M.M.; et al. Rat Lungworm Infection Associated with Central Nervous System Disease—Eight U.S. States, January 2011–January 2017. *MMWR Morb. Mortal. Wkly. Rep.* **2018**, *67*, 825–828. [CrossRef]

5. Johnston, D.I.; Dixon, M.C.; Elm, J.L.; Calimlim, P.S.; Sciulli, R.H.; Park, S.Y. Review of Cases of Angiostrongyliasis in Hawaii, 2007–2017. *Am. J. Trop. Med. Hyg.* **2019**, *101*, 608–616. [CrossRef] [PubMed]
6. Murphy, G.S.; Johnson, S. Clinical Aspects of Eosinophilic Meningitis and Meningoencephalitis Caused by *Angiostrongylus cantonensis*, the Rat Lungworm. *Hawaii J. Med. Public Health* **2013**, *72*, 35–40.
7. Barratt, J.; Chan, D.; Sandaradura, I.; Malik, R.; Spielman, D.; Lee, R.; Marriott, D.; Harkness, J.; Ellis, J.; Stark, D. *Angiostrongylus cantonensis*: A Review of Its Distribution, Molecular Biology and Clinical Significance as a Human Pathogen. *Parasitology* **2016**, *143*, 1087. [CrossRef]
8. Lunn, J.; Lee, R.; Martin, P.; Malik, R. Antemortem Diagnosis of Canine Neural Angiostrongylosis Using ELISA. *Aust. Vet. J.* **2003**, *81*, 128–131. [CrossRef]
9. Ansdell, V.; Kramer, K.J.; McMillan, J.K.; Gosnell, W.L.; Murphy, G.S.; Meyer, B.C.; Blalock, E.U.; Yates, J.; Lteif, L.; Smith, O.A.; et al. Guidelines for the Diagnosis and Treatment of Neuroangiostrongyliasis: Updated Recommendations. *Parasitology* **2021**, *148*, 227–233. [CrossRef]
10. Qvarnstrom, Y.; Xayavong, M.; da Silva, A.C.A.; Park, S.Y.; Whelen, A.C.; Calimlim, P.S.; Sciulli, R.H.; Honda, S.A.A.; Higa, K.; Kitsutani, P.; et al. Real-Time Polymerase Chain Reaction Detection of *Angiostrongylus cantonensis* DNA in Cerebrospinal Fluid from Patients with Eosinophilic Meningitis. *Am. J. Trop. Med. Hyg.* **2016**, *94*, 176–181. [CrossRef]
11. Sears, W.J.; Qvarnstrom, Y.; Dahlstrom, E.; Snook, K.; Kaluna, L.; Baláž, V.; Feckova, B.; Šlapeta, J.; Modry, D.; Jarvi, S.; et al. AcanR3990 QPCR: A Novel, Highly Sensitive, Bioinformatically-Informed Assay to Detect *Angiostrongylus cantonensis* Infections. *Clin. Infect. Dis.* **2021**, *73*, e1594–e1600. [CrossRef] [PubMed]
12. Wilkins, P.P.; Qvarnstrom, Y.; Whelen, A.C.; Saucier, C.; da Silva, A.J.; Eamsobhana, P. The Current Status of Laboratory Diagnosis of *Angiostrongylus cantonensis* Infections in Humans Using Serologic and Molecular Methods. *Hawaii J. Med. Public Health* **2013**, *72*, 55–57. [PubMed]
13. Cross, J.H. Clinical Manifestations and Laboratory Diagnosis of Eosinophilic Meningitis Syndrome Associated with Angiostrongyliasis. *Southeast Asian J. Trop. Med. Public Health* **1978**, *9*, 161–170. [PubMed]
14. Chen, S.N. Enzyme-Linked Immunosorbent Assay (ELISA) for the Detection of Antibodies to *Angiostrongylus cantonensis. Trans. R. Soc. Trop. Med. Hyg.* **1986**, *80*, 398–405. [CrossRef] [PubMed]
15. Kum, W.W.S.; Ko, R.C. Isolation of Specific Antigens from *Angiostrongylus cantonensis* by a Combination of Preparative Isoelectric Focusing and Affinity Chromatography. *Trop. Biomed.* **1986**, *3*, 55–60.
16. Yen, C.M.; Chen, E.R. Detection of Antibodies to *Angiostrongylus cantonensis* in Serum and Cerebrospinal Fluid of Patients with Eosinophilic Meningitis. *Int. J. Parasitol.* **1991**, *21*, 17–21. [CrossRef]
17. Eamsobhana, P.; Yoolek, A.; Suvouttho, S.; Suvouttho, S. Purification of a Specific Immunodiagnostic *Parastrongylus cantonensis* Antigen by Electroelution from SDS-Polyacrylamide Gels. *Southeast Asian J. Trop. Med. Public Health* **2001**, *32*, 6.
18. Morassutti, A.L.; Rascoe, L.N.; Handali, S.; DA Silva, A.J.; Wilkins, P.P.; Graeff-Teixeira, C. Cross-Reactivity of the 31 KDa Antigen of *Angiostrongylus cantonensis*—Dealing with the Immunodiagnosis of Meningoencephalitis. *Parasitology* **2017**, *144*, 459–463. [CrossRef]
19. Yong, W.K.; Dobson, C. Antibody Responses in Rats Infected with *Angiostrongylus cantonensis* and the Passive Transfer of Protective Immunity with Immune Serum. *Z. Parasitenkd.* **1982**, *67*, 329–336. [CrossRef]
20. Takai, A.; Sato, Y.; Otsuru, M. Polyclonal B Cell Activation during the Course of *Angiostrongylus cantonensis* Infection. *Dev. Comp. Immunol.* **1985**, *9*, 485–495. [CrossRef]
21. Kanbara, T.; Ohmomo, N.; Umemura, T.; Kajigaya, H.; Umeda, M.; Sugiyama, M.; Isoda, M. Local Antibody Production and Immune Complex Formation in Rats Experimentally Infected with *Angiostrongylus cantonensis. Am. J. Trop. Med. Hyg.* **1988**, *39*, 353–360. [CrossRef] [PubMed]
22. Chang, C.; Chao, D. Time-Course Changes in Antibody Responses to Different Stages of *Angiostrongylus cantonensis* in Experimentally Infected Rats. *Zool. Stud.* **1996**, *35*, 62–67.
23. Dorta-Contreras, A.J.; Reiber, H. Intrathecal Synthesis of Immunoglobulins in Eosinophilic Meningoencephalitis Due to *Angiostrongylus cantonensis. Clin. Diagn. Lab. Immunol.* **1998**, *5*, 452–455. [CrossRef] [PubMed]
24. Jarvi, S.I.; Cannoles, K.; Eamsobhana, P.; Kaluna, L.M.; Shepherd, L.; Tagami, Y. Immunodiagnostic Detection of *Angiostrongylus cantonensis* Exposure on Hawaii Island Using Isogeographical 31 KDa Antigen. *Am. J. Trop. Med. Hyg.* 2023, *in press*.
25. Jarvi, S.I.; Jacob, J.; Sugihara, R.T.; Leinbach, I.L.; Klasner, I.H.; Kaluna, L.M.; Snook, K.A.; Howe, M.K.; Jacquier, S.H.; Lange, I.; et al. Validation of a Death Assay for *Angiostrongylus cantonensis* Larvae (L3) Using Propidium Iodide in a Rat Model (*Rattus norvegicus*). *Parasitology* **2019**, *146*, 1421–1428. [CrossRef]
26. Jarvi, S.I.; Eamsobhana, P.; Quarta, S.; Howe, K.; Jacquier, S.; Hanlon, A.; Snook, K.; McHugh, R.; Tman, Z.; Miyamura, J.; et al. Estimating Human Exposure to Rat Lungworm (*Angiostrongylus cantonensis*) on Hawaii Island: A Pilot Study. *Am. J. Trop. Med. Hyg.* **2020**, *102*, 69–77. [CrossRef]
27. Eamsobhana, P.; Mak, J.W.; Yong, H.S. Identification of *Parastrongylus cantonensis* Specific Antigens for Use in Immunodiagnosis. *Int. Med. Res. J.* **1997**, *1*, 1–4.
28. Bustin, S.A.; Benes, V.; Garson, J.A.; Hellemans, J.; Huggett, J.; Kubista, M.; Mueller, R.; Nolan, T.; Pfaffl, M.W.; Shipley, G.L.; et al. The MIQE Guidelines: Minimum Information for Publication of Quantitative Real-Time PCR Experiments. *Clin. Chem.* **2009**, *55*, 611–622. [CrossRef]

29. Schnyder, M.; Jefferies, R.; Schucan, A.; Morgan, E.R.; Deplazes, P. Comparison of Coprological, Immunological and Molecular Methods for the Detection of Dogs Infected with *Angiostrongylus Vasorum* before and after Anthelmintic Treatment. *Parasitology* **2015**, *142*, 1270–1277. [CrossRef]
30. Odani, J.S.; Cowie, R.H.; Malik, R.; Jarvi, S.I. Rat Lungworm Disease for the Veterinary Professional. *CTAHR Vet. Ext.* **2019**, *VE-2*, 1–9.

pathogens

Viewpoint

Neuroangiostrongyliasis: Updated Provisional Guidelines for Diagnosis and Case Definitions

Carlos Graeff-Teixeira [1,*], Kittisak Sawanyawisuth [2], Shan Lv [3], William Sears [4], Zhaily González Rodríguez [5], Hilda Hernández Álvarez [5], Pedro Casanova Arias [5], Leticia Karolini Walger Schultz [1], Alicia Rojas [6], John Jacob [7], Susan Jarvi [7] and Kenton Kramer [8,†] on behalf of International Network on Angiostrongyliasis—INA

1 Center for Health Sciences, Universidade Federal do Espírito Santo, Vitória 29075-910, ES, Brazil; leticia.walgers@gmail.com
2 Department of Medicine, Faculty of Medicine, Khon Kaen University, Khon Kaen 40000, Thailand; yong2001th@yahoo.com
3 Chinese Center for Disease Control and Prevention, National Institute of Parasitic Diseases, Shanghai 200025, China; lvshan@nipd.chinacdc.cn
4 Laboratory of Parasitic Diseases, National Institutes of Health, Bethesda, MD 20892, USA; william.sears@nih.gov
5 Reference National Parasitology Laboratory, Instituto de Medicina Tropical "Pedro Kourí" (IPK), Havana 11400, Cuba; zhaily@ipk.sld.cu (Z.G.R.); hilda@ipk.sld.cu (H.H.Á.); rcapedro@ipk.sld.cu (P.C.A.)
6 Departamento de Parasitologia, Facultad de Microbiologia, Universidad de Costa Rica, San Jose 11501-2060, Costa Rica
7 Department of Pharmaceutical Sciences, Daniel K. Inouye College of Pharmacy, University of Hawaii at Hilo, Hilo, HI 96720, USA
8 Department of Tropical Medicine, Medical Microbiology and Pharmacology, John A. Burns School of Medicine, University of Hawaii at Manoa, Honolulu, HI 96822, USA
* Correspondence: graeff.teixeira@gmail.com; Tel.: +55-51-999818599
† All INA members that are coauthors are mentioned in the Acknowledgements.

Abstract: *Angiostrongylus cantonensis* is the main causative agent for eosinophilic meningoencephalitis in humans. Larvae are rarely found in the cerebral spinal fluid (CSF). Consequently, serology and DNA detection represent important diagnostic tools. However, interpretation of the results obtained from these tools requires that more extensive accuracy studies be conducted. The aim of the present study is to update guidelines for diagnosis and case definitions of neuroangiostrongyliasis (NA) as provided by a working group of a recently established International Network on Angiostrongyliasis. A literature review, a discussion regarding criteria and diagnostic categories, recommendations issued by health authorities in China and an expert panel in Hawaii (USA), and the experience of Thailand were considered. Classification of NA cases and corresponding criteria are proposed as follows: minor (exposure history, positive serology, and blood eosinophilia); major (headache or other neurological signs or symptoms, CSF eosinophilia); and confirmatory (parasite detection in tissues, ocular chambers, or CSF, or DNA detection by PCR and sequencing). In addition, diagnostic categories or suspected, probable, and confirmatory are proposed. Updated guidelines should improve clinical study design, epidemiological surveillance, and the proper characterization of biological samples. Moreover, the latter will further facilitate accuracy studies of diagnostic tools for NA to provide better detection and treatment.

Keywords: angiostrongyliasis; neuroangiostrongyliasis; rat lungworm disease; eosinophilic meningitis

check for updates

Citation: Graeff-Teixeira, C.; Sawanyawisuth, K.; Lv, S.; Sears, W.; Rodríguez, Z.G.; Álvarez, H.H.; Arias, P.C.; Schultz, L.K.W.; Rojas, A.; Jacob, J.; et al. Neuroangiostrongyliasis: Updated Provisional Guidelines for Diagnosis and Case Definitions. *Pathogens* **2023**, *12*, 624. https://doi.org/10.3390/pathogens12040624

Academic Editor: Fabrizio Bruschi

Received: 21 March 2023
Revised: 6 April 2023
Accepted: 14 April 2023
Published: 20 April 2023

1. Introduction

Angiostrongylus cantonensis is an intra-arterial nematode which, in accidental human hosts, can cause eosinophilic meningitis [1]. Eosinophilic inflammatory responses in the central nervous system (CNS) are mainly elicited by helminths. However, these inflammatory responses can also occur in association with cancer, intra-thecal drugs, and intra-vesicular devices [2]. The non-infectious causes for eosinophilic meningoencephalitis are dominant in many regions, e.g., Central Europe.

A. cantonensis are parasites that live inside the pulmonary arteries of rodents, and mollusks serve as intermediate hosts. Several other invertebrates, such as shrimp, frogs, and lizards, can serve as paratenic hosts. Larvae of *A. cantonensis* develop in the fibromuscular tissues of mollusks, and subsequently infect humans when raw or undercooked food is ingested [1,3]. Larvae-contaminated water may also be a source of human infection [4,5]. In accidental hosts, including humans, larvae migrate and are retained in the CNS. This retention prevents complete maturation of adult worms inside the pulmonary arteries [1,3].

Larvae are rarely detected in examinations of cerebrospinal fluid (CSF), thereby preventing confirmation of a diagnosis by direct identification of parasites [6]. Therefore, immunological and DNA detection methods are important tools for evaluating patients with suspected neuroangiostrongyliasis (NA) [7–9]. While antibody and DNA-detection methods have been standardized, evaluations of these methods as reliable detection tools have been less than adequate due to the small number of well-characterized biological reference samples that are currently available from different geographical areas. More recently, a highly sensitive and specific quantitative PCR method has been developed to confirm diagnosis of NA [10].

A. cantonensis is native to southeastern Asia and the Pacific Islands [11], yet its presence has expanded to multiple continents. This perceived expansion may also have resulted from an increased awareness and availability of diagnostic capability. It is also a food-borne disease that has been linked to travelers. A rough estimate of the cumulative number of reported cases worldwide is 2800 [3]. Despite the potential for severe CNS disease, NA is not a highly prevalent infection worldwide. To date, the Hawaiian Islands and southeastern Asia, especially Thailand, have been the most affected endemic areas. However, reduced rates of occurrences have been observed in recent years [12,13].

Disease caused by *A. cantonensis* includes isolated meningeal lesions and meningitis associated with brain tissue inflammation (meningoencephalitis) [14]. More rarely, ocular angiostrongyliasis can develop [15]. For severe cases of encephalitis, lethality may reach 80% [16]. The objective of the present work is to update and explore a possible consensus regarding diagnostic criteria and case definitions for NA. It is anticipated that such effort will improve patient management worldwide, will promote comparable clinical and epidemiological studies, and will define conditions for establishing an international biobank of well characterized samples. The latter would represent a valuable resource for evaluations of diagnostic tests to achieve detection and control of NA.

2. Materials and Methods

Several meetings of the International Symposium on Angiostrongyliasis (ISA), also named the International Symposium on Rat lungworm Disease, have been held in various countries over many years, namely, Thailand, China, Hawaii (USA), and Australia. The Canary Islands will host the next ISA in September 2023. When a group of researchers and clinician attendees of the ISA met in an online discussion in October 2021, the need for more extensive studies, including multicenter accuracy studies, of diagnostic tests for NA was highlighted. Basic requirements for such studies are: (i) clear and well-defined diagnostic criteria; and (ii) establishment of a collaborative international biobank with well-characterized biological samples. On 23 November 2021, the International Network on Angiostrongyliasis was established. Subsequently, several online meetings were hosted throughout 2022 to discuss many issues, including diagnostic criteria. Health authorities in China and an expert panel in Hawaii (USA) have independently established recommendations for diagnosing and treating NA [17,18]. Diagnostic and patient management experience from Thailand, currently the most endemic country, are also available. In particular, the results from a systematic review published by Khamsai and collaborators (2020) were examined [13]. Based on these considerations, the following revised guidelines were developed. They are not to be definitive but represent a starting point for continuous improvement.

3. Results

The present updated guidelines were developed based on the following principles: (i) to provide clear definition and expression ("taxonomy") of symptoms, signs, or laboratory results; (ii) to select the most relevant symptoms most closely related to NA (sensitivity) and classify them as minor, major, or confirmatory criteria (meanwhile, excessive value for general, less specific symptoms and signs is avoided); and (iii) to define categories of diagnosis according to the degree of certainty for etiological diagnosis, from lower (suspected) to highest (probable and confirmed) degrees, conforming to standard organization of diagnostic and treatment guidelines [19–23].

Elements of exposure history are summarized in Table 1. Criteria are classified as minor, major, or confirmatory (Table 2 and Figure 1). Diagnosis categories (suspected, probable, and confirmed) and recommended actions are also presented in Table 3 and Figure 1. Antibody detection (serology), blood eosinophilia, and history of exposure are considered minor criteria, since their isolated presence does not constitute strong evidence for NA. In addition, these criteria may be absent in patients. While serological studies are useful for epidemiological exposure studies [5,24], cross-reactivity and persistence of antibodies after cure are recognized as universal limitations of serology for the detection of current infections.

Table 1. *Angiostrongylus cantonensis* and elements from exposure history with increased risk of transmission, according to Wang et al. [3], Khamsai et al. [25], Ansdell et al. [18], and Howe et al. [4].

Type of Exposure	Vectors/ Transmission Areas
Ingestion of raw, undercooked and/or inadequately washed foods	Mollusks: snails, slugs Salads Juices Fruits Planarians Freshwater shrimp Crabs Frogs Lizards Water contaminated with larvae
Touching, handling	Mollusks, snails, or slugs
Residence or recent travel	Endemic areas

Table 2. Diagnostic criteria for neuroangiostrongyliasis and criteria classification according to the strength of evidence for the etiological diagnosis.

Criteria Category	Criteria
Minor	a. Exposure history b. Serology (antibody detection) c. Blood eosinophilia
Major	a. CSF [1] eosinophilia b. Headache, other neurological signs or symptoms, and other obvious etiologies ruled out.
Confirmatory	a. Larvae in tissues, CSF, or eye b. DNA detection [2]

[1] CSF: cerebrospinal fluid. [2] Antigen detection in CSF (no test currently available) may be an additional criterion for confirmed diagnosis in the future (see Sears et al., for novel highly sensitive PCR [10]).

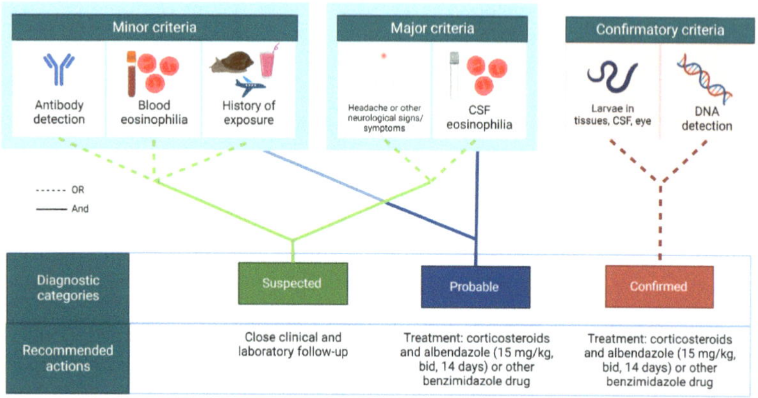

Figure 1. Diagnostic criteria classification, case definitions, and recommendations for follow up and treatment of patients with neuroangiostrongyliasis, as proposed by a working group from the International Network on Angiostrongyliasis.

Table 3. Classification of diagnostic categories for neuroangiostrongyliasis and recommended actions: follow up and treatment.

Diagnosis Categories	Criteria	Recommended Actions
Suspected	• Headache, OR other neurological signs/symptoms, OR • CSF eosinophilia [1] AND • Any minor criteria [2]	Close clinical and laboratory follow up
Probable	• Headache, OR other neurological signs/symptoms AND • CSF eosinophilia [1] AND • At least two minor criteria [2]	[3] Treatment: corticosteroids and albendazole (15 mg/kg, bid, 14 days)
Confirmed	• Larvae in tissues, CSF, or eye chambers OR • DNA detection [4]	[3] Treatment: corticosteroids and albendazole (15 mg/kg, bid, 14 days)

If serology is positive [1] and CSF eosinophilia [2] is higher than 40%, consider NA as highly probable/suspected. [3] For details and treatment alternatives, see Sawanyawisuth, K. and Sawanyawisuth [14] and Jacob et al. [26]. [4] Antigen detection in CSF (no test currently available) may be an additional criterion for confirmed diagnosis in the future (see Sears et al., for novel highly sensitive PCR [10]).

Major criteria, headache, or other neurological signs or symptoms associated with CSF eosinophilia are proposed as defining criteria for eosinophilic meningitis or meningoencephalitis. Importantly, these major criteria are labeled as such to be highly suggestive for NA, since *A. cantonensis* is their main causative agent. Clinical manifestations' strength of evidence for NA depends on the absence of other obvious causes for eosinophilic meningitis or meningoencephalitis [18].

Finding parasitic structures in cerebrospinal fluid or brain tissues is extremely rare yet represents undisputed criteria for confirming a diagnosis. Ocular examination, including anterior and posterior chambers, may also disclose larvae or even adult worms since the eye is an area of localization second to meningeal vessels for *A. cantonensis* [15]. Worms may rarely be found inside pulmonary arteries in fatal cases [27–29].

4. Discussion

In 2006, the Chinese Ministry of Health published recommendations on diagnostic criteria for angiostrongyliasis and proposed three case definitions: (i) suspected; (ii) clinical, and (iii) parasitologically diagnosed [17]. Eating history (C1), clinical manifestations (C2), blood eosinophilia (C3), CSF eosinophilia (C4), seropositivity (C5), and presence of parasites in CSF or other sites (C6) were the suggested criteria. Suspected NA is considered whenever a combination of two criteria of C1 to C5 is observed. A clinical diagnosis is considered when C1, C2, C3, and C4 are present. NA is confirmed with demonstration of parasites (C6).

In 2018, a State Task Force in Hawaii (USA) prepared guidelines for diagnosis and treatment of NA. An update was subsequently published in 2020 [18]. The latter guidelines consider two diagnostic categories: presumptive and definitive, or confirmed by identification of parasites or DNA detection with PCR. A presumptive diagnosis is established by: (i) characteristic symptoms and signs, (ii) exposure history, and (iii) CSF eosinophilia.

Headache is the most common clinical manifestation and can occur in the absence of other neurological signs or symptoms. Neck stiffness and fever are also present in 15% and 40% of cases, respectively, and occur more often in children [26,30]. Among less common manifestations, dysesthesias (paresthesias and hyperesthesias) and migratory myalgia may be valuable to indicate neuroangiostrongyliasis, which are to be properly investigated in prospective clinical studies [3,6]. Other early symptoms can be prodromal symptoms; that is, those due to the physiological and neurological damages due to the larval migration from the gastrointestinal tract into the CNS [31,32]. It is likely that the severity of prodromal symptoms is directly associated with the number of parasites involved in the infection. Larvae attempting to penetrate the gastrointestinal walls may cause symptoms such as nausea, vomiting, abdominal pain, or diarrhea. Larval migration through the liver, kidneys, and lungs may cause malaise, low-grade fever, coughs, jaundice-like symptoms, and hematuria. Larvae stranded beneath the skin may produce rashes or pruritis-like symptoms. Prodromal symptoms are nonspecific; thus, unless there is a great degree of suspicion of infection, it is unlikely to alert the medical practitioner [32,33].

CSF eosinophilia is another main indicator for NA and occurs in approximately 50% of patients [26]. To date, the definition of "CSF eosinophilia" remains controversial. Some authors consider any number of eosinophils as abnormal, while other authors have selected 10% or an absolute number of 10 eosinophils as a threshold [18,26]. The degree of eosinophilia present may be a stronger indicator for NA, and patients with $\geq$40% are more likely to have NA [34]. In some cases, only a follow-up lumbar puncture can reveal CSF eosinophilia [3]. However, lack of appropriate staining and differential counting of CSF cells can prevent demonstration of an eosinophilic inflammatory response in meningeal tissues and fluids. Physicians are urged to check with their laboratory to determine if a proper examination was performed.

Different transmission areas may present predominant intentional or nonintentional exposure behavior. For example, in Thailand, intentional food habits favor transmission; while in Hawaii, ingestion of contaminated food or water is usually non-intentional. More severe cases in Hawaii may be due to higher intake of larvae because of the high burden of infection of terrestrial gastropods. A definite exposure history may be absent [18]. The precise date of exposure is important to consider since the incubation period (IP) for NA is usually between 1 and 3 weeks, although a 1-year IP has been reported [3,16,32]. In addition to the well-known role of several foods as a source for infection, there are indications that larvae can be ingested in drinking water (Table 1) [4]. Thus, knowledge of active transmission areas may help increase and sustain awareness regarding NA, help identify cases for early treatment, and promote the prevention of more severe disease. In endemic areas, knowledge and attention to non-specific clinical manifestations (e.g., fever, nausea, vomiting, agitation, lethargy) can facilitate early diagnosis and treatment. There is need for continued research to decrease the time to diagnosis. Perhaps antigen capture assays for blood, stool, or urine will be helpful. Post-exposure treatment with pyrantel has been shown to be promising, although its clinical relevance is yet to be demonstrated, especially considering the very short time frame from exposure to the effective prevention of larvae migration through intestinal mucosa [35].

While several serological tests have been developed, limitations involving cross-reactivity [36,37], late seroconversion [38], and less than adequate accuracy evaluations have prevented these results from being confirmatory. Correspondingly, in the present classification proposed, positive serological examination is recognized as a minor criterion. Seroconversion will be helpful to confirm the diagnosis when CSF cannot be collected and an accurate diagnostic test is available.

The proposed guidelines advocate for early treatment of "probable" cases (without other obvious causes), as well as confirmed NA cases (Table 2). A possible exception for early treatment recommendation is prompt recovery (less than 24 h) from headache and other neurological deficits without any clinical manifestations suggestive of encephalitic compromise. Prospective studies are needed to confirm and optimize diagnostic workflow and case definitions. Corticosteroids are a cornerstone of NA management since they can potentially reduce both the intensity and duration of headaches, the main cause of distress in patients [39–41]. Non-steroid anti-inflammatory drugs should not be administered along with corticosteroids because of increased risk for upper gastrointestinal bleeding [39]. Measures to reduce intracranial pressure, such as therapeutic repeated CSF removal, have been shown to be effective, and are a choice symptomatic treatment for alleviating severe headaches in patients affected by NA [39]. Benzimidazole anthelmintic drugs, especially albendazole, are also recommended despite multiple controversies regarding their safety and efficacy [42]. For detailed discussion and recommendations for angiostrongyliasis treatment, see specific reports and reviews [14,25,41].

5. Conclusions

In conclusion, it is anticipated that the present global revision and updated recommendations for diagnosis and treatment of NA will facilitate much needed clinical studies, with the use of standardized diagnostic criteria leading to better comparative studies of patients from different geographic areas. The present revisions and updates, intended to support further discussions and developments, may also help provide well-defined diagnostic categories for public health surveillance and a strategy for better characterizing biological samples for accuracy studies of diagnostic tools. The latter is especially relevant for evaluations of newly developed methods for early and specific detection and treatment of NA.

Author Contributions: Conceptualization, C.G.-T.; Discussion and analysis: K.S., S.L., W.S., Z.G.R., H.H.Á., P.C.A., A.R., J.J., S.J. and K.K.; Methodology and Data Curation, L.K.W.S.; Writing—Original Draft Preparation, C.G.-T. and A.R.; Writing—Review and Editing, C.G.-T., K.S., S.L., W.S., Z.G.R., H.H.Á., P.C.A., A.R., J.J., S.J. and K.K.; Coordination, C.G.-T. All authors have read and agreed to the published version of the manuscript.

Funding: This research received no external funding.

Institutional Review Board Statement: Not applicable.

Informed Consent Statement: Not applicable.

Data Availability Statement: All data is presented in the paper.

Acknowledgments: Authors acknowledge the support from INA coordinators and hosts of the next International Symposium on Angiostrongyliasis (https://ratlungworm2023tenerife.com/), The Canary Islands, September 2023: Claudia Paredes (claudia.paredes@uib.es Universitat de les Illes Balears) and Pilar Foronda (pforonda@ull.edu.es, Universidad de La Laguna), Spain. The present provisional guidelines are not official statements from any public health organization or academic institution.

Conflicts of Interest: The authors declare no conflict of interest.

References

1. Alicata, J.E. Biology and distribution of the rat lungworm, *Angiostrongylus cantonensis*, and its relationship to eosinophilic meningoencephalitis and other neurological disorders of man and animals. *Adv. Parasitol.* **1965**, *3*, 223–248. [PubMed]
2. Graeff-Teixeira, C.; Silva, A.C.A.; Yoshimura, K. Update on eosinophilic meningoencephalitis and its clinical relevance. *Clin. Microbiol. Rev.* **2009**, *22*, 322–348. [CrossRef] [PubMed]
3. Wang, Q.P.; Lai, D.H.; Zhu, X.Q.; Chen, X.G.; Lun, Z.R. Human angiostrongyliasis. *Lancet Infect. Dis.* **2008**, *8*, 621–630. [CrossRef]
4. Howe, K.; Kaluna, L.; Lozano, A.; Torres Fischer, B.; Tagami, Y.; McHugh, R.; Jarvi, S. Water transmission potential of *Angiostrongylus cantonensis*: Larval viability and effectiveness of rainwater catchment sediment filters. *PLoS ONE* **2019**, *14*, e0209813. [CrossRef]
5. Jarvi, S.I.; Eamsobhana, P.; Quarta, S.; Howe, K.; Jacquier, S.; Hanlon, A.; Snook, K.; McHugh, R.; Tman, Z.; Miyamura, J.; et al. Estimating Human Exposure to Rat Lungworm (*Angiostrongylus cantonensis*) on Hawai'i Island: A Pilot Study. *Am. J. Trop. Med. Hyg.* **2020**, *102*, 69–77. [CrossRef]
6. Punyagupta, S.; Juttijudata, P.; Bunnag, T. Eosinophilic meningitis in Thailand. Clinical studies of 484 typical cases probably caused by *Angiostrongylus cantonensis*. *Am. J. Trop. Med. Hyg.* **1975**, *24*, 921–931. [CrossRef]
7. Eamsobhana, P.; Yong, H.S. Immunological diagnosis of human angiostrongyliasis due to *Angiostrongylus cantonensis* (Nematoda: Angiostrongylidae). *Int. J. Infect. Dis.* **2009**, *13*, 425–431. [CrossRef]
8. Qvarnstrom, Y.; Xayavong, M.; Silva, A.C.A.; Park, S.Y.; Whelen, A.C.; Calimlim, P.S.; Silva, A.J. Real-time polymerase chain reaction detection of *Angiostrongylus cantonensis* DNA in cerebrospinal fluid from patients with eosinophilic meningitis. *Am. J. Trop. Med. Hyg.* **2016**, *94*, 176. [CrossRef]
9. Somboonpatarakun, C.; Intapan, P.M.; Sadaow, L.; Rodpai, R.; San pool, O.; Maleewong, W. Development of an immunochromato graphic device to detect antibodies for rapid diagnosis of human angiostrongyliasis. *Parasitology* **2020**, *147*, 194–198. [CrossRef]
10. Sears, W.J.; Qvarnstrom, Y.; Dahlstrom, E.; Snook, K.; Kaluna, L.; Baláž, V.; Feckova, B.; Šlapeta, J.; Modry, D.; Jarvi, S.; et al. AcanR3990 qPCR: A Novel, Highly Sensitive, Bioinformatically-Informed Assay to Detect *Angiostrongylus cantonensis* Infections. *Clin. Infect. Dis.* **2021**, *73*, e1594–e1600. [CrossRef]
11. Kliks, M.M.; Palumbo, N.E. Eosinophilic meningitis beyond the Pacific basin: The global dispersal of a peridomestic zoonosis caused by *Angiostrongylus cantonensis*, the nematode lungworm of rats. *Soc. Sci. Med.* **1992**, *34*, 199–212. [CrossRef] [PubMed]
12. Johnston, D.I.; Dixon, M.C.; Elm, J.L., Jr.; Calimlim, P.S.; Sciulli, R.H.; Park, S.Y. Review of cases of angiostrongyliasis in Hawaii, 2007–2017. *Am. J. Trop. Med. Hyg.* **2019**, *101*, 608. [CrossRef] [PubMed]
13. Khamsai, S.; Chotmongkol, V.; Tiamkao, S.; Maleewong, W.; Limpawattana, P.; Boonsawat, W.; Sawanyawisuth, K. Eosinophilic meningitis epidemiological data from a national database in Thailand's Department of Disease Control: A pragmatic, retrospective analytical study. *Orphanet J. Rare Dis.* **2022**, *17*, 1–6. [CrossRef] [PubMed]
14. Sawanyawisuth, K.; Sawanyawisuth, K. Treatment of angiostrongyliasis. *Trans. R. Soc. Trop. Med. Hyg.* **2008**, *102*, 990–996. [CrossRef]
15. Sawanyawisuth, K.; Kitthaweesin, K.; Limpawattana, P.; Intapan, P.; Tiamkao, S.; Jitpimolmard, S.; Chotmongkol, V. Intraocular angiostrongyliasis: Clinical findings, treatments and outcomes. *Trans. R. Soc. Trop. Med. Hyg.* **2007**, *101*, 497–501. [CrossRef] [PubMed]
16. Chotmongkol, V.; Sawanyawisuth, K. Clinical manifestations and outcome of patients with severe eosinophilic meningoencephalitis presumably caused by *Angiostrongylus cantonensis*. *Southeast Asian J. Trop. Med. Public Health* **2002**, *45*, 30–61.
17. Lv, S.; Zhang, Y.; Chen, S.-R.; Wang, L.-B.; Fang, W.; Chen, F.; Jiang, J.-Y.; Li, Y.-L.; Du, Z.-W.; Zhou, X.-N. Human Angiostrongyliasis Outbreak in Dali, China. *PLoS Negl. Trop. Dis.* **2009**, *3*, e520. [CrossRef]
18. Ansdell, V.; Kramer, K.J.; McMillan, J.K.; Gosnell, W.L.; Murphy, G.S.; Meyer, B.C.; Blalock, E.U.; Yates, J.; Leteif, L.; Smith, O.A.; et al. Guidelines for the diagnosis and treatment of neuroangiostrongyliasis: Updated recommendations. *Parasitology* **2021**, *148*, 227–233. [CrossRef]
19. Dechet, A.M.; Parsons, M.; Rambaran, M.; Mohamed-Rambaran, P.; Florendo-Cumbermack, A.; Persaud, S.; Baboolal, S.; Ari, M.D.; Shadomy, S.V.; Zaki, S.R.; et al. Leptospirosis Outbreak following Severe Flooding: A Rapid Assessment and Mass Prophylaxis Campaign; Guyana, January–February 2005. *PLoS ONE* **2012**, *7*, e39672. [CrossRef]
20. OPAS/WHO. Modulos de Principios de Epidemiología para el Control de Enfermedades. In *Modulo 4: Vigilancia em Salud Publica*; Organização Pan-Americana da Saúde: Brasília, Brazil, 2010; 52p.
21. Colt, S.; Garcia-Casal, M.N.; Peña-Rosas, J.P.; Finkelstein, J.L.; Rayco-Solon, P.; Weise, Z.C.; Mehta, S. Transmission of Zika virus through breast milk and other breastfeeding-related bodily-fluids: A systematic review. *PLoS Negl. Trop. Dis.* **2017**, *11*, e0005528. [CrossRef]
22. WHO. Yellow fever surveillance and outbreak response: Revision of case definitions, October 2010. *Wkly. Epidemiol. Rec.* **2010**, *47*, 19.
23. WHO. Surveillance, Case Investigation and Contact Tracing for Mpox (Monkeypox): Interim Guidance, 22 December 2022. *WHO/MPX/Surveillance/2022.4*. Available online: https://www.who.int/publications/i/item/WHO-MPX-Surveillance-2022.4 (accessed on 26 January 2023).
24. Jarvi, S.I.; Nakayama, K.; Eamsobhana, P.; Kaluna, L.; Shepherd, L.; Tagami, Y. Immunodiagnostic detection of *Angiostrongylus cantonensis* exposure on Hawai'i Island using isogeographical 31 kDa antigen. *Am. J. Trop. Med. Hyg.* **2023**, in press.

25. Khamsai, S.; Chindaprasirt, J.; Chotmongkol, V.; Tiamkao, S.; Limpawattana, P.; Senthong, V.; Sawanyawisuth, K. Clinical features of eosinophilic meningitis caused by *Angiostrongylus cantonensis* in Thailand: A systematic review. *Asia-Pac. J. Sci. Technol.* **2020**, *25*, APST-25-02-09. Available online: https://so01.tci-thaijo.org/index.php/APST/article/view/240510 (accessed on 20 March 2023).
26. Jacob, J.; Steel, A.; Howe, K.; Jarvi, S.I. Management of Rat Lungworm Disease (Neuroangiostrongyliasis) Using Anthelmintics: Recent Updates and Recommendations. *Pathogens* **2023**, *12*, 23. [CrossRef] [PubMed]
27. Sonakul, D. Pathological findings in four cases of human angiostrongyliasis. *Southeast Asian J. Trop. Med. Public Health* **1978**, *9*, 220–227. [PubMed]
28. Cooke-Yarborough, C.M.; Kornberg, A.J.; Hogg, G.G.; Spratt, D.M.; Forsyth, J.R. A fatal case of angiostrongyliasis in an 11-month-old infant. *Med. J. Aust.* **1999**, *170*, 541–543. [CrossRef]
29. Prociv, P.; Carlisle, M.S. The spread of *Angiostrongylus cantonensis* in Australia. *Southeast Asian J. Trop. Med. Public Health* **2001**, *32*, 126–128.
30. Sawanyawisuth, K.; Chindaprasirt, J.; Senthong, V.; Limpawattana, P.; Auvichayapat, N.; Tassniyom, S.; Chotmongkol, V.; Maleewong, W.; Intapan, P.M. Clinical manifestations of Eosinophilic meningitis due to infection with *Angiostrongylus cantonensis* in children. *Korean J. Parasitol.* **2013**, *51*, 735–738. [CrossRef]
31. Sawanyawisuth, K.; Pugkhem, A.; Mitchai, J.; Intapan, P.M.; Anunnatsiri, S.; Limpawattana, P.; Chotmongkol, V. Abdominal angiostrongyliasis caused by *Angiostrongylus cantonensis*: A possible cause of eosinophilic infiltration in human digestive tract. *Pathol. Res. Pract.* **2010**, *206*, 102–104. [CrossRef]
32. Yii, C.Y. Clinical observations on eosinophilic meningitis and meningoencephalitis caused by *Angiostrongylus cantonensis* inTaiwan. *Am. J. Trop. Med. Hyg.* **1976**, *25*, 233–249. [CrossRef]
33. Cross, J.H. Clinical manifestations and laboratory diagnosis of eosinophilic meningitis syndrome associated with angiostrongyliasis. *Southeast Asian J. Trop. Med. Public Health* **1978**, *9*, 161–170. [PubMed]
34. Sawanyawisuth, K.; Sawanyawisuth, K.; Senthong, V.; Limpawattana, P.; Phichaphop, A.; Intapan, P.M.; Maleewong, W.; Tiamkao, S.; Jitpimolmard, S.; Chotmongkol, V. How can clinicians ensure the diagnosis of meningitic angiostrongyliasis? *Vector-Borne Zoonotic Dis.* **2012**, *12*, 73–75. [CrossRef] [PubMed]
35. Jacob, J.; Steel, A.; Kaluna, L.; Hess, S.; Leinbach, I.; Antaky, C.; Sugihara, R.; Hamilton, L.; Follett, P.; Howe, K.; et al. In vivo efficacy of pyrantel pamoate as a post-exposure prophylactic for rat lungworm (*Angiostrongylus cantonensis*). *Int. J. Parasitol. Drugs Drug Resist.* **2022**, *19*, 1–5. [CrossRef]
36. Cognato, B.B.; Handali, S.; Morassutti, A.L.; Silva, A.J.; Graeff-Teixeira, C. Heterologous expression of three antigenic proteins from *Angiostrongylus cantonensis*: ES-7, Lec-5, and 14-3-3 in mammalian cells. *Mol. Biochem. Parasitol.* **2018**, *221*, 32–35. [CrossRef] [PubMed]
37. Morassutti, A.L.; Levert, K.; Perelygin, A.; Silva, A.J.; Wilkins, P.; Graeff-Teixeira, C. The 31-kDa antigen of *Angiostrongylus cantonensis* comprises distinct antigenic glycoproteins. *Vector-Borne Zoonotic Dis.* **2012**, *12*, 961–968. [CrossRef] [PubMed]
38. Murphy, G.S.; Johnson, S. Clinical aspects of eosinophilic meningitis and meningoencephalitis caused by *Angiostrongylus cantonensis*, the rat lungworm. *Hawaii J. Med. Public Health* **2013**, *72*, 35.
39. Sawanyawisuth, K.; Limpawattana, P.; Busaracome, P.; Ninpaitoon, B.; Chotmongkol, V.; Intapan, P.M.; Tanawirattananit, S. A 1-week course of corticosteroids in the treatment of eosinophilic meningitis. *Am. J. Med.* **2004**, *117*, 802–803. [CrossRef]
40. Ansdell, V.; Wattanagoon, Y. *Angiostrongylus cantonensis* in travelers: Clinical manifestations, diagnosis, and treatment. *Curr. Opin. Infect. Dis.* **2018**, *31*, 399–408. [CrossRef]
41. Khamsai, S.; Sawanyawisuth, K.; Senthong, V.; Limpawattana, P.; Chindaprasirt, J.; Intapan, P.M.; Ngamjarus, C. Corticosteroid treatment reduces headache in eosinophilic meningitis: A systematic review. *Drug Target Insights* **2021**, *15*, 1. [CrossRef]
42. Jacob, J.; Steel, A.; Lin, Z.; Berger, F.; Zöeller, K.; Jarvi, S.I. Clinical Efficacy and Safety of Albendazole and Other Benzimidazole Anthelmintics for Rat Lungworm Disease (Neuroangiostrongyliasis): A Systematic Analysis of Clinical Reports and Animal Studies. *Clin. Infect. Dis.* **2022**, *74*, 1293–1302. [CrossRef]

pathogens

MDPI

Article

Successful Removal of *Angiostrongylus cantonensis* Larvae from the Central Nervous System of Rats 7- and 14-Days Post-Infection Using a Product Containing Moxidectin, Sarolaner and Pyrantel Embonate (Simparica Trio™) in Experimental Infections

Michaela Henry [1], Richard Malik [1,2,*], Jan Šlapeta [1] and Rogan Lee [3]

1 Sydney School of Veterinary Science, Faculty of Science, The University of Sydney, Sydney, NSW 2006, Australia
2 Centre for Veterinary Education, The University of Sydney, Sydney, NSW 2006, Australia
3 The University of Sydney Institute for Infectious Diseases, Westmead Hospital, The University of Sydney, Sydney, NSW 2006, Australia
* Correspondence: richard.malik@sydney.edu.au

Abstract: *Angiostrongylus cantonensis* is a nematode with an indirect lifecycle, using molluscs as intermediate hosts. Rats are the definitive host. By administering a suitable anthelmintic, at an appropriate interval, the risk of clinical neuroangiostrongyliasis occurring in paratenic hosts (e.g., dogs, man) can be eliminated. We wanted to determine if infective larvae (L3) of *A. cantonensis* can be safely killed during their migration through the central nervous system (CNS) by oral administration of an anthelmintic combination containing moxidectin (480 μg/kg, Simparica Trio™; M-S-P), thereby preventing patent infections in rats. Eighteen rats were used: ten received oral M-S-P every four weeks; eight rats were used as controls. Rats were initially given M-S-P as a chew to eat, but an acquired food aversion meant that subsequent doses were given by orogastric lavage. All 18 rats were challenged once or twice with approximately 30 L3 *A. cantonensis* larvae via orogastric lavage. Infection status was determined by faecal analysis using the Baermann technique and necropsy examination of the heart, pulmonary arteries and lungs. Eight out of ten rats dosed with M-S-P had zero lungworms at necropsy; a single female worm was detected in each of the remaining two rats. No treated rats had L1 larvae in faeces. In contrast, all eight controls were infected with patent infections, with a median of 14.5 worms per rat detected at necropsy. The difference in infection rates was significant (two tailed Fishers Exact; $p = 0.0011$). Moxidectin given orally once every month killed migrating larvae before they reached the pulmonary arteries in 80% of treated rats, while in 20%, only a single female worm was present. Considering the short half-life of moxidectin in the rat, it is likely that the effectiveness of moxidectin is due to larvicidal action on migrating L3, L4 and L5 larvae in the brain parenchyma or subarachnoid space, either 7 days (L3/L4 in cerebrum and spinal cord) or 14 days (L4/L5 in cerebrum and subarachnoid space) after inoculation. This study is a prelude for future research to determine if monthly moxidectin administration orally as M-S-P could prevent symptomatic neuroangiostrongyliasis in dogs.

Keywords: *Angiostrongylus cantonensis*; rat lungworm; moxidectin; Simparica

check for updates

Citation: Henry, M.; Malik, R.; Šlapeta, J.; Lee, R. Successful Removal of *Angiostrongylus cantonensis* Larvae from the Central Nervous System of Rats 7- and 14-Days Post-Infection Using a Product Containing Moxidectin, Sarolaner and Pyrantel Embonate (Simparica Trio™) in Experimental Infections. *Pathogens* **2023**, *12*, 305. https://doi.org/10.3390/pathogens12020305

Academic Editor: Susan Jarvi

Received: 13 December 2022
Revised: 1 February 2023
Accepted: 9 February 2023
Published: 12 February 2023

1. Introduction

The rat lungworm *Angiostrongylus cantonensis* is a nematode with a complex indirect life cycle. This involves an unusual obligatory migration stage, in which L3 larvae migrate through the central nervous system (CNS), where they grow, moult and further mature. The parasite uses molluscs, such as snails and slugs, as intermediate hosts, while rats are the definitive host [1]. Rats and paratenic hosts become infected by ingesting L3 larvae present in tissues of the intermediate and paratenic (transport) hosts. *A. cantonensis* usually does not produce overt clinical signs in rats, unless the definitive host ingests a heavy dose of infective larvae over a short time. If this happens, verminous pneumonia develops, clinically manifested by dyspnoea and reduced exercise tolerance [1–4]. Rarely, transient neurological signs occur; although, usually despite extensive neural migration by larvae, neurological signs in rats are absent or unappreciated.

Adult lungworms live in the pulmonary arteries of several rat species. The life cycle is complex, involving snails as intermediate hosts, and a variety of different species as paratenic hosts. Male and female lungworm mate in the pulmonary arteries. Females lay eggs which embolise to the pulmonary parenchyma. After hatching, first stage L1 larvae move up the mucociliary escalator, are swallowed, and appear in faecal pellets of the rats, which are attractive to snails. Snails become infected by ingestion of infective first stage larvae, which develop and mature into L3 larvae. The life cycle is completed when rats ingest snails. L3 larvae penetrate the intestine of the rat, travel in the portal circulation to the liver, and then rapidly reach the systematic circulation. The most unusual aspect of the lifecycle involves an obligatory migration of L3 larvae in the rat central nervous system (CNS). Larvae migrate widely through the CNS, including the spinal cord, brain and optic nerves, growing, moulting and growing further until they are ready to leave the CNS via the arachnoid villi to make their way back to the right ventricle and pulmonary arteries [1–4].

In contradistinction to the situation in rats, the migration of infective L3 larvae can cause serious disease in accidental hosts such as dogs, wildlife (birds, possums, bats) and humans [1–5]. Rat lungworm disease (neuroangiostrongyliasis) is most often manifested as eosinophilic meningoencephalitis [3,4], and peripheral eosinophilia is often reported in canine and human dead-end hosts [2]. The actual migration of larvae through the CNS and the inflammatory response thereby incited can together give rise to hindlimb weakness, paralysis and even death, if the animal does not receive treatment. This is due to severe damage to the CNS, especially the spinal cord and cauda equina, and in some cases the brain [4–6].

To minimise the prevalence of clinical rat lungworm disease, it is necessary to interrupt critical portions of the life cycle of *A. cantonensis*, thereby limiting the number of infections, and their extent, in dogs, wildlife and man [6–9]. Although it is certainly helpful to reduce the number of rats and snails in the environment, the strategy most likely to be successful at preventing canine infections is the administration of prophylactic anthelmintics. The key consideration for anthelmintic choice is the half-life of the drug and its dosing frequency.

The macrocyclic lactone moxidectin has a long half-life in most species, including dogs. When given at monthly intervals, it has been shown to be highly effective in preventing infection of dogs with *Angiostrongylus vasorum* [10,11], a closely related parasite of dogs which has a similar lifecycle, but without the novel larval migration though the CNS. It is therefore to be expected that moxidectin, if given at the same dosage interval, is also likely to prevent infective larvae of *A. cantonensis* reaching and damaging the CNS of dogs [12,13]. This is because L3 larvae migrating through canine tissues on their way to the CNS will be killed in transit if moxidectin concentrations are sufficiently high in the blood and extracellular fluid, which is likely for up to a month after dosing with systemic moxidectin, as a result of its long half-life [14,15].

The aim of this study was to evaluate the efficacy of an orally administered combination of sarolaner/moxidectin/pyrantel embonate (Simparica Trio™ [M-S-P]; Zoetis; [16]) given every four weeks against subsequent challenge with *A. cantonensis* L3 larvae in rats. When we designed this experiment, the pharmacokinetics of moxidectin in the rat had not yet been determined, and we erroneously thought that the half-life in the rat would be comparable to that in the dog (see later).

2. Materials and Methods

2.1. Animals

Eighteen juvenile male rats (*Rattus norvegicus*; Wistar strain) were used for this study. All rats were approx. 4–5 months-of-age at the start of the study; Group 1 rats were 19 weeks, Group 2 rats were 14 weeks, while Group 3 and 4 rats were 11 weeks old at the start of the experiment. The median weight of the rats at the start of the experiment was 505 g. Rats were housed in pairs in cages with an appropriate substrate (wood shavings), enrichment items (e.g., cardboard rolls), and were fed commercial rat cubes and provided with fresh tap water ad libitum.

The first dose of M-S-P was given to the rats in tablet form. Our assumption was that a product developed to be palatable chew for dogs, would also be palatable for rats, and a pilot dose given to rats (not part of this study cohort) was consumed. Food was withheld from each treatment rat overnight to ensure the subjects were hungry. Each rat was isolated in their own cage the following day with no bedding and just the half tablet, with rats monitored over the course of the day. This proved satisfactory on the first occasion that the product was provided. However, when given the second and third doses of M-S-P, rats were reluctant to eat the 'chew' voluntarily, probably because they had developed a taste aversion. To circumvent this, the half tablet was dissolved in 3 mL of distilled water to form a suspension and was administered to each rat via orogastric lavage under light isoflurane anaesthesia. Rats were monitored shortly after the procedure for any evidence of aspiration, regurgitation or vomiting.

All 18 rats were each challenged, on one or two occasions, with approx. 30 *A. cantonensis* L3 larvae. Infective L3 larvae were first freshly harvested from macerated tissues of chronically infected snails 2–3 h prior to administration. Larvae were administered to rats via oral gavage using a plastic pipette under light isoflurane anaesthesia. One of the investigators (RL) counted out 30 larvae as they were sucked up into a pipette tip; the approximate final volume was made up with distilled water to a volume of 500 μL. These larvae were then instilled into the distal oesophagus, although it was not possible to control for the loss of some larvae (via subsequent regurgitation and vomiting) during the gavage process. We expected that 50–75% of administered larvae would reach the stomach due to losses of larvae during this procedure.

The control group consisted of eight rats, four of which were given 30 L3 at 2 weeks (Group 1), and four of which were given the same dose at 7 weeks (Group 2). The treatment group consisted of ten rats, divided into two groups of five, that had been given M-S-P orally at zero weeks, four weeks and eight weeks by voluntary intake (t = 0), and subsequently using gastric lavage (t = 4 and 8 weeks) (Table 1). The monthly administration of M-S-P was chosen to mimic the situation whereby pet dogs are given this product on an ongoing monthly basis, while potentially being exposed to rat lungworm larvae at random occasions.

One treatment group (Group 3) was challenged with infectious L3 larvae only once (at week 2; two weeks after M-S-P), while the other treatment group (Group 4) was challenged twice (at week 2 and week 7; two and three weeks after M-S-P, respectively).

Table 1. Timetable that outlines the schedule of when rats were challenged with *A. cantonensis* L_3 larvae and when the Simparica Trio™ (M-S-P) dose was administered.

Week	Group 1 (Control; n = 4)	Group 2 (Control; n = 4)	Group 3 (ST; n = 5)	Group 4 (ST; n = 5)
0			M-S-P	M-S-P
1				
2	30 L_3 larvae PO		30 L_3 larvae PO	30 L_3 larvae PO
3				
4			M-S-P	M-S-P
5				
6				
7		30 L_3 larvae PO		30 L_3 larvae PO
8			M-S-P	M-S-P
9				
10				
11				
12				
13				
14	Necropsy	Necropsy	Necropsy	Necropsy

Orange font highlights infective dose of L3 larvae; Blue highlights dosing with moxidectin in Simparica Trio; the blue arrows indicate the infective larvae that are targeted by the moxidectin in M-S-P given 14 or 7 days later.

2.2. Dosage Calculation

Rats received sarolaner/moxidectin/pyrantel (half of the Simparica Trio™ tablet, 10.1–20 kg size; Zoetis, Sydney, New South Wales, Australia). This tablet contains 24 mg sarolaner, 480 μg of moxidectin and 100 mg pyrantel [16]. Therefore, each rat voluntarily consumed or was gavaged with approximately 240 μg of moxidectin. For a rat weighing 500 g, this equates to a dose of 480 μg/kg. The dose calculation assumes that the tablet is homogenous in its formulation, which is unproven.

2.3. Examination for Patent Infection and Presence of *A. cantonensis* L1 in Rat Faeces

The Baermann technique was used to extract *A. cantonensis* L1 in rat faeces [17,18]. A wet preparation slide with a coverslip was made and viewed using conventional light microscopy using the 10× objective lens. Larvae in faeces were highly motile when detected.

2.4. Necropsy Examination of Rats to Detect Adults *A. cantonensis* in the Right Ventricle and Pulmonary Artery of Rats

All rats were humanely euthanised at week 14 by inspiration of 100% carbon dioxide. Each rat was weighed using electronic scales immediately after euthanasia. The heart and lungs were removed from the chest cavity by a combination of blunt and sharp dissection to locate all adult nematodes in the right ventricle and/or pulmonary arteries. The numbers of male and female worms were determined by examination under a dissecting microscope. The worms were small, ranging from approximately 15–25 mm in length with the females having a slightly larger diameter and longer length, as well as the 'barber's pole' appearance, caused by the alimentary tract (containing digested blood) and reproductive tract being wrapped around each other [2,4,9].

2.5. Statistical Analysis

The weights of treated and control rats were compared using the Mann–Whitney U test. A two-tailed Fisher's Exact test was used to compare the number of adult *A. cantonensis* worms present in treated versus control rats. The two M-S-P treatment groups (of 5 rats) were combined (10 rats in total) and compared to the 8 control rats.

3. Results

There were no mortalities or treatment-related adverse reactions over the course of the study. All rats continued to grow and increase in body mass during the experiment. None of the rats appeared dyspneic (at rest) at any time during the experiment.

Both control groups, Group 1 and Group 2, were successfully infected with *A. cantonensis* L3 larvae (Table 2). A total of seven out of eight control rats had adult worms (11–23) present on necropsy at Week 14 and all seven were positive for *A. cantonensis* L1 larvae on Baermann examination of faecal pellets. One of the eight control rats (from Group 1) had no *A. cantonensis* L1 larvae in its faeces and only a single *A. cantonensis* adult male worm present in a pulmonary artery. Considering that approx. 30 L3 were given to each rat, the resulting worm burden (median 14.5 worms; IQR 11.5 to 16.5 worms) per rat was consistent with about half of the larvae reaching maturity. Noticeable lesions in the pulmonary parenchyma were present in all the control rats except for the rat with the single worm infection (Figure 1). The gross pulmonary lesions were mostly localised in the caudodorsal portions of the lungs.

Table 2. Pooled results of the worm burden found in the control and treatment groups. Rats were dosed with approx. 30 L_3 larvae.

Parasite Status	Group 1 (Control; *n* =4) One Challenge at 2 Weeks	Group 2 (Control; *n* = 4) One Challenge at 7 Weeks	Group 3 (M-S-P; *n* = 5) One Challenge at 2 Weeks	Group 4 (M-S-P; *n* = 5) Two Challenges at 2 and 7 Weeks
Infected	4/4	4/4	0/5	2/5
***A.c.*Male**	4,5,10,1	17,8,10,6	0,0,0,0,0	0,0,0,0,0
***A.c.*Female**	8,6,6,0	6,9,6,7	0,0,0,0,0	1,0,0,0,1
***A.c.*Total**	12,11,16,1	23,17,16,13	0,0,0,0,0	1,0,0,0,1
L_1 larvae in fresh faeces	3/4 rats positive (1 rat with a single male worm was negative)	4/4 positive	0/5 positive	0/5 positive (2/5 rats each had a single female worm and were negative)
Lung lesions	3/4	4/4	0/5	0/5
Body weight (g)	690,674,662,595 (median 668)	580,673,595,591 (median 593)	646,586,595,616,538 (median 595)	584,565,633,616,589 (median 589)

A.c. Angiostrongylus cantonensis.

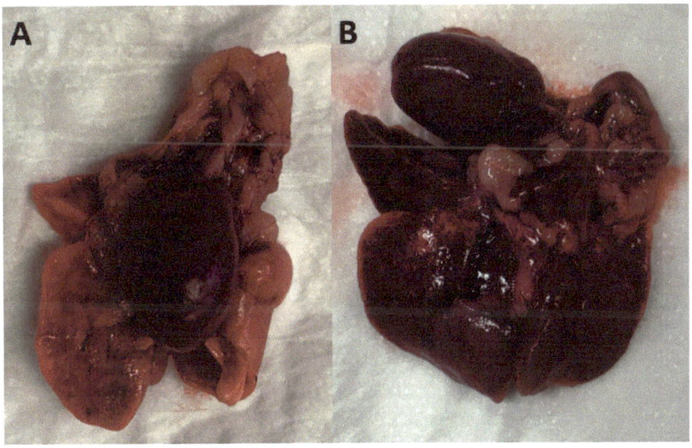

Figure 1. Heart and lungs dissected from Wistar rats at the end of the experiment ex vivo. In (**A**), a rat treated with moxidectin in M-S-P has normal heart and lungs, while a control rat with a moderate burden of mature *A. cantonensis* is shown in (**B**) on the right.

All four Group 3 rats that were treated with M-S-P on weeks 0, 4 and 8 weeks remained negative despite a single challenge with *A. cantonensis* L3 larvae at week 2 (Table 1). Group 4 rats were treated on weeks 0, 4 and 8 and challenged with *A. cantonensis* L3 larvae at week 2 (2 weeks after M-S-P) and again at week 7 (3 weeks after M-S-P; Table 1). All four rats were negative for *A. cantonensis* L1 larvae in their faeces, but at necropsy, two out of five rats each had a single female *A. cantonensis* recovered from the pulmonary arteries.

Thus, 8/10 rats dosed with M-S-P had zero lungworms at necropsy; in the remaining two rats, which might not have received a full dose of M-S-P for technical reasons, a single female worm was detected. In contrast, 8/8 control rats were infected with *A. cantonensis*, with a median of 14.5 worms per rat detected at necropsy and patent infections with motile L1 abundant in faecal pellets. The difference in infection rates was highly significant (two tailed Fishers Exact; $p = 0.0011$; [19]).

We can also express the results in terms of percentage burden reduction. In control rats, 109 worms were present in 8 rats; whereas in rats given M-S-P monthly, only 2 worms were present in 10 rats; so, the burden reduction was 13.625 worms per rat to 0.2 worms per rat, which is a 98.5% reduction. Monthly M-S-P administration prevents the shedding of L1 larvae in all rats, regardless of whether there was one or two challenges with infective larvae. Likewise, monthly M-S-P prevented the development any of discernible gross lung pathology at necropsy examination.

There was no significant difference between the weights of the treated and control rats at the end of the experiment (Mann–Whitney U test; U = 22; $p = 0.12$; [20]).

4. Discussion

M-S-P is a fixed dose combination of three anthelmintic drugs designed to be administered monthly in dogs to prevent heartworm disease, tick paralysis, flea, lice and mite infestations, and intestinal nematode infections [16]. Its spectrum covers all important and common helminth infections except tapeworms (cestodes). The experiments described here represent a model (pilot experiments) for a conceptually similar study that we hope to undertake in dogs to determine if it is possible to prevent them from getting neuroangiostrongyliasis when given M-S-P as a monthly preventative.

In this study, rats were used as a surrogate for dogs because it is much easier to obtain animal ethics approval for rat experiments in our jurisdiction and such trials are substantially less expensive. Our aim was to use the moxidectin component of M-S-P to interrupt the life cycle by killing migrating L3 larvae in the CNS, thereby preventing them maturing into adult nematodes and reproducing within the definitive host [3,9]. Because M-S-P is given either 1 week or 2 weeks after larval challenge, the L3 larvae have all left the gut and entered the CNS. Furthermore, because the L3 are no longer in the gut, the pyrantel in the M-P-S does not contribute any effect on larvae, as pyrantel does not achieve effective concentrations in the CNS, which is where the L3 larvae are at this point in time. Finally, sarolaner is an isoxazoline ectoparasiticide thought to have no effect on nematode larvae, but is rather a selective inhibitor of arthropod γ-aminobutyric acid- and l-glutamate-gated chloride channels in fleas and ticks.

What we would actually like to show in dogs is that moxidectin concentrations in plasma are sufficiently high to kill L3 before they enter the CNS, but this is harder to prove in rats, as the migration of modest larval burdens do not usually cause observable neurologic signs. In rats, the L3 larvae can be found circulating in blood within a few hours of inoculation, and within 24 h, L3 larvae have entered the spinal cord and brain where they grow, moult twice, approximately on days 7 and 14 post-infection, then reach the subarachnoid space from where they eventually leave the CNS as L5 larvae, on the way to the right ventricle and pulmonary arteries [3,4,21].

The pharmacokinetics of moxidectin given orally to Wistar rats had not been determined at the start of the research project, which commenced in mid-February 2021 during the COVID pandemic. It was presumed that moxidectin would have a half-life in rats comparable to what has been reported in the dog, such that monthly administration would have substantial and cumulative activity, such that when rats would be challenged with an oral dose of infective larvae, blood concentrations of moxidectin might be sufficient to kill the larvae before they migrate to the CNS. Based on this notion, the experimental protocol set out in Table 1 was constructed.

However, with the benefit of the data from Buchter and colleagues published in July 2021, serum levels of moxidectin are essentially nil by 48 h after oral administration of 500–750 μg/kg [22]. The half-life of moxidectin given orally to dogs is very long viz. 621 h (13.9 to 25.9 days), while in Wistar rats the half-life is only 10.4 h; this means that therapeutic blood levels of moxidectin are unlikely to be maintained for more than 2–3 days in our rats, whereas in dogs the coverage extends to approximately a full calendar month [14,15,22]. In other words, if M-S-P is given at t = 0, 4 weeks and 8 weeks, then blood concentrations would have fallen to zero by the time the inoculum of 30 L3 larvae were given at week 2 or week 7.

Therefore, the action of moxidectin in M-S-P is attributable to its larvicidal action on migrating L3, L4 and L5 larvae in the CNS when it is given at 4 weeks and 8 weeks, with the M-S-P reaching short-lived therapeutic levels 2 weeks or 1 week, respectively, after larval challenge (Figure 2). Thus, our results confirm those of Schmahl et al. [18] who showed that moxidectin given transdermally at 4–32 mg/kg (with imidacloprid as the topical Advocate™; Elanco) at 15 days post-infection was highly effective at killing the 'CNS-dwelling larvae' of *A. cantonensis* [3,4,18,21]. It is, in some respects, remarkable that the death and disintegration in the order of 11–23 larvae (expected to be approx. 3 mm long [21]; Figure 2) did not produce more discernible neurological signs that might be appreciated even by a short daily examination.

In our study, M-S-P was administered either 7 or 14 days ***after*** challenge with infective larvae. Larvae are in the CNS at this stage and remain susceptible to moxidectin, as this macrocyclic lactone readily crosses the blood brain barrier (BBB) [23,24]. Furthermore, inflammation from the migrating larvae may have caused the BBB to become leaky and pro-inflammatory cytokines are known to inhibit the p-glycoprotein pump, allowing more of the lipophilic drug to enter the extracellular fluid around the parasite [23,24]. This means that in the first rat treatment group (Group 3), it was the M-S-P containing moxidectin given 2 weeks after larval challenge which killed larvae migrating through the CNS, when the moxidectin blood and CNS concentrations were sufficiently high. The same was true in the first challenge of the second treatment group of rats (Group 4), but in the second challenge, the M-S-P containing moxidectin given 1 week after larval challenge was the dose that was effective. It was fortuitous that the dose of moxidectin we selected based on the older literature and allometric scaling was similar to doses informed by recent pharmacokinetic studies. The chosen dose of approx. 480 μg/kg was close to the most efficacious dose (500 μg/kg) used for treating *Strongyloides ratti* infections in rats [22].

As stated, this study was conducted in the spirit of being a pilot experiment, as there had been limited previous research into the chemoprophylaxis of *A. cantonensis* infection in rats or dogs. The product M-S-P was selected as it represents the drug combination of greatest potential to prevent the important parasitic diseases of companion dogs in eastern Australia, including rat lungworm disease. The 'palatable chew' formulation of M-S-P, however, contributed to dosage inaccuracy, in that we do not know whether the active was uniformly distributed in the tablet, nor were we sure that rats swallowed all of the drug when it was given by gavage under light gaseous anaesthesia. This may have contributed to the two rats with a single worm infection despite monthly M-S-P administration. Although rats would eat the tablet once, they would not eat them subsequently, probably because of the development of a food aversion to one or more of the active ingredients. It was thus necessary to create a drug suspension administered by gastric lavage under anaesthesia for subsequent administrations. The half a tablet dose was not fully administered, as a small amount of residual solution was always left in the dead space of the pipette. It would have been much easier if the rats could have consumed the complete dose on their own.

Presumably, this would not be a problem if the experiments were repeated in dogs, the species for which the palatable chew was developed.

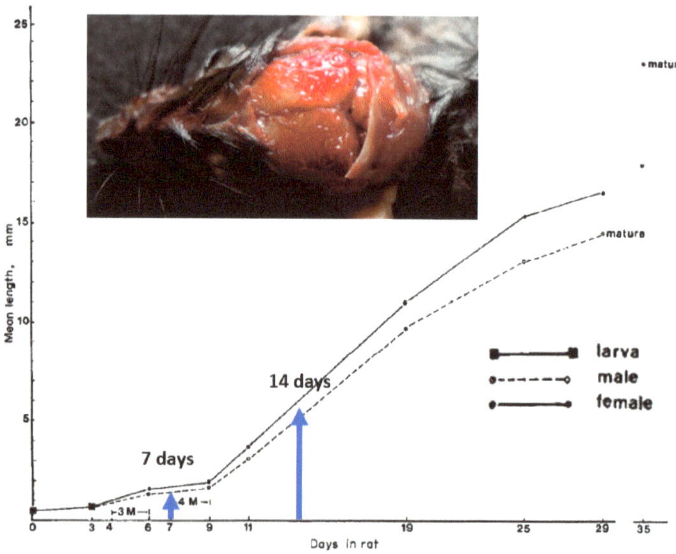

Figure 2. Growth of *A. cantonensis* in experimental white Wistar rats; 3 M = 3rd moult; 4 M = 4th moult. The blue arrows indicate the size of the migrating larvae at 7 days and 14 days when they would be exposed to moxidectin following the administration of Simparica Trio™ by ingestion following gastric lavage. Inset: photo of a wild rat after removal of the calvarium to demonstrate young adult worms (L5) in the subarachnoid space overlying the cerebrum. Photograph courtesy of Lydia Tong and Derek Spielman, Taronga Zoo; original diagram adapted from Manoon Bhaibulaya's classic paper in 1973 [25], with modifications.

A further complication of the oral gavage technique also applies to administering the infective L3 larvae. The exact number of L3 larvae administered orally to each rat was variable from rat to rat, which would in part explain some of the variation in worm numbers observed in control rats at necropsy. When performing oral gavage, a portion of the solution will always remain inside the dead space of the pipette. Thus, a few L3 larvae probably remained in the pipette after each administration. For this reason, a new pipette was used for each gavage. In relation to the single control rat that had a much lower worm burden than the rest of the controls, it is suspected that the oral gavage of the L3 larvae was not technically perfect in this individual. This control rat did develop a single worm infection, but the worm burden was smaller than the other controls, suggesting most larvae were not swallowed, or were swallowed and then regurgitated. Indeed, a risk of oral gavage under gaseous anaesthesia is that gastric reflux and aspiration pneumonia can occur with poor technique and especially with insufficiently deep anaesthesia [25].

Moxidectin is an anthelmintic used to prevent other nematode species such as the heartworm *Dirofilaria immitis* [12] and in the UK and Europe, *Angiostrongylus vasorum* [11,13,26]. This latter parasite causes 'French heartworm', a complicated disease resulting from the presence of many adult worms in the pulmonary arteries of infected dogs. This canine disease can be prevented by the monthly administration of moxidectin, which kills both migrating larvae and adult worms in the pulmonary arteries. The efficacy of this drug is in part due to its very long half-life in the dog, which results in cumulative kinetics with effective blood concentrations of moxidectin present for the entire month when the product is given every four weeks. Moxidectin targets nematodes by opening chloride channels in their cell membranes, causing lethal paralysis [27]. Based on the results presented here and in the work of Schmahl and colleagues [10], the moxidectin in M-S-P provided blood and CNS concentrations sufficient to prevent rat lungworm developing to patency following an oral challenge with L3 larvae. Interestingly, Schmahl et al. [10] observed that the doses of moxidectin they used had no effect on adult lungworms in the pulmonary artery of rats, but given the recent work showing the short half-life of moxidectin in the rat [12], it seems most likely that mature worms require more than 1–2 days of effective moxidectin blood concentrations to succumb.

Moxidectin is generally considered superior to other macrocyclic lactone anthelmintics such as ivermectin, selamectin and milbemycin, with moxidectin showing a faster onset of activity and a greater efficacy, for example against L3 larvae of *Strongyloides* spp. [22]. Moxidectin also has a substantially longer half-life and a greater area under the curve when given orally to dogs compared to when it is given to rats [14–16,22]. Moxidectin remains in the blood for sufficiently long in the dog that the drug accumulates when given monthly, resulting in progressively higher serum and tissue concentrations. Such prolonged high levels likely explain its efficacy against mature *A. vasorum* worms in the pulmonary arteries of dogs [14,15].

In the dog, the prolonged kinetics of moxidectin would likely prevent neuroangiostrongyliasis by killing infective L3 larvae before they would reach the CNS, a mechanism not observed in these experimental rats because of the much shorter half-life of moxidectin in this species. However, this needs to be confirmed experimentally. Moxidectin in various formulations is already being used prophylactically with this intention in highly endemic areas, such as in Hawaii, Sydney, Brisbane and along the east coast of Australia. A further issue relates to whether the long-acting depot formulations of moxidectin (Proheart SR12™; Zoetis) would produce sufficiently high concentration in serum to kill infective third stage larvae of *A. cantonensis*, and for how long [14].

Completion of this study has helped to further expand the knowledge of anthelmintics used in rats. Monthly moxidectin would prevent pet rats from developing a patent *A. cantonensis* infection should they eat an infected slug or snail.

5. Conclusions

This study demonstrates unequivocally that administering Simparica Trio™ containing moxidectin can interrupt the lifecycle of *A. cantonensis* in rats by causing lethal paralysis of infective larvae migrating through the spinal cord, peripheral nerves, brain and subarachnoid space. This research conducted on rats can be used to help guide preventive care in domestic animals such as dogs, as well as wildlife and zoo animals in high prevalence areas with endemic *A. cantonensis*.

Author Contributions: Conceptualisation, M.H., R.M., J.Š. and R.L.; methodology, M.H., R.M., J.Š. and R.L; formal analysis, M.H., R.M., J.Š. and R.L.; investigation M.H., R.M., J.Š. and R.L; resources, R.L.; data curation, R.L.; writing—M.H.; writing—review and editing, M.H., R.M., J.Š. and R.L; supervision, R.M., J.Š. and R.L.; project administration, R.L.; funding acquisition, J.Š. All authors have read and agreed to the published version of the manuscript.

Funding: This study was funded by Richard Malik, Rogan Lee and student (DVM) research funds provided by the Sydney School of Veterinary Science, University of Sydney.

Institutional Review Board Statement: The animal study protocol was approved by the Animal Ethics Committee of Westmead Hospital, NSW Government Western Sydney Local Health District (approval number was 5167.12.20 and date of approval).

Informed Consent Statement: Not applicable.

Data Availability Statement: All the data applicable to this investigation are presented in this manuscript.

Acknowledgments: The authors appreciate the assistance of the on-site personnel at Westmead vivarium who cared for these rats over the course of study. We also wish to thank the veterinarian who oversees the vivarium, Ross Matthews, for advice about ethics and welfare of the rats and the conduct of the experiment.

Conflicts of Interest: The authors declare no conflict of interest.

References

1. Traub, R.; Brown, G.; Coleman, G.; Constantinoiu, C.; O'Handley, R.; Slapeta, J. Parasites of the dog and cat. 4.1.4. Parasites causing multi-systemic disease in small animals: *Angiostrongylus* spp. (*Parastrongylus*). In *Australian Animal Parasites Inside & Out*; Brown, G., Coleman, G., Constantinoiu, C., Gasser, R., Holyoake, P., Hobbs, R., Lymbery, A., O'Handley, R., Phalen, D., Pomroy, W., et al., Eds.; The Australian Society for Parasitology: Cairns North, Australia, 2015; pp. 546–549.
2. Wun, M.K.; Davies, S.; Spielman, D.; Lee, R.; Hayward, D.; Malik, R. Gross, microscopic, radiologic, echocardiographic and haematological findings in rats experimentally infected with *Angiostrongylus cantonensis*. *Parasitology* **2020**, *148*, 159–166. [CrossRef] [PubMed]
3. Barratt, J.; Chan, D.; Sandaradura, I.; Malik, R.; Spielman, D.; Lee, R.; Marriott, D.; Harkness, J.; Ellis, J.; Stark, D. *Angiostrongylus cantonensis*: A review of its distribution, molecular biology and clinical significance as a human pathogen. *Parasitology* **2016**, *143*, 1087–1118. [CrossRef] [PubMed]
4. Mackerras, M.; Sandars, D. The life history of the rat lung-worm, *Angiostrongylus cantonensis* (Chen) (Nematoda: Metastrongylidae). *Aust. J. Zool.* **1955**, *3*, 1–21. [CrossRef]
5. Lee, R.; Pai, T.-Y.; Churcher, R.; Davies, S.; Braddock, J.; Linton, M.; Yu, J.; Bell, E.; Wimpole, J.; Dengate, A.; et al. Further studies of neuroangiostrongyliasis (rat lungworm disease) in Australian dogs: 92 new cases (2010–2020) and results for a novel, highly sensitive qPCR assay. *Parasitology* **2020**, *148*, 178–186. [CrossRef] [PubMed]
6. Mason, K.V. Canine neural angiostrongylosis: The clinical and therapeutic features of 55 natural cases. *Aust. Veter. J.* **1987**, *64*, 201–203. [CrossRef] [PubMed]
7. Lunn, J.A.; Lee, R.; Smaller, J.; MacKay, B.M.; King, T.; Hunt, G.B.; Martin, P.; Krockenberger, M.B.; Spielman, D.; Malik, R. Twenty two cases of canine neural angiostrongylosis in eastern Australia (2002–2005) and a review of the literature. *Parasites Vectors* **2012**, *5*, 70. [CrossRef] [PubMed]
8. Garcia, J.S.; Lúcio, C.D.S.; Bonfim, T.C.D.S.; Junior, A.M.; Tunholi, V.M.; Tunholi-Alves, V.M.; Mota, E.M.; Simões, R.D.O.; Santana, A.C.; Hooper, C.; et al. Metabolic and histopathological profile of *Rattus norvegicus* (Wistar) experimentally infected by *Angiostrongylus cantonensis* (Chen, 1935). *Exp. Parasitol.* **2014**, *137*, 35–40. [CrossRef] [PubMed]
9. Cowie, R.H. Biology, systematics, life cycle, and distribution of *Angiostrongylus cantonensis*, the cause of rat lungworm disease. *Hawai'i J. Med. Public Health A J. Asia Pac. Med. Public Health* **2013**, *72* (Suppl. S2), 6–9.
10. Willesen, J.; Kristensen, A.T.; Jensen, A.; Heine, J.; Koch, J. Efficacy and safety of imidacloprid/moxidectin spot-on solution and fenbendazole in the treatment of dogs naturally infected with *Angiostrongylus vasorum* (Baillet, 1866). *Veter. Parasitol.* **2007**, *147*, 258–264. [CrossRef]
11. Becskei, C.; Willesen, J.L.; Schnyder, M.; Wozniakiewicz, M.; Miroshnikova, N.; Mahabir, S.P. Field safety and efficacy of an orally administered combination of sarolaner, moxidectin and pyrantel (Simparica Trio®) for the prevention of angiostrongylosis in dogs presented as veterinary patients. *Parasites Vectors* **2020**, *13*, 385. [CrossRef]
12. Kryda, K.; Six, R.H.; Walsh, K.F.; Holzmer, S.J.; Chapin, S.; Mahabir, S.P.; Myers, M.; Inskeep, T.; Rugg, J.; Cundiff, B.; et al. Laboratory and field studies to investigate the efficacy of a novel, orally administered combination product containing moxidectin, sarolaner and pyrantel for the prevention of heartworm disease (*Dirofilaria immitis*) in dogs. *Parasites Vectors* **2019**, *12*, 445. [CrossRef] [PubMed]
13. Di Cesare, A.; Traversa, D. Canine angiostrongylosis: Recent advances in diagnosis, prevention, and treatment. *Vet. Med.* **2014**, *5*, 181–192. [CrossRef]
14. The European Agency for the Evaluation of Medical Products; Committee for Veterinary Medicinal Products. Moxidectin. 1997. Available online: https://www.ema.europa.eu/en/documents/mrl-report/moxidectin-summary-report-1-committee-veterinary-medicinal-products_en.pdf (accessed on 20 May 2021).
15. Bowman, D.D.; Charles, S.D.; Arther, R.G.; Settje, T. Laboratory Evaluation of the Efficacy of 10% Imidacloprid + 2.5% Moxidectin Topical Solution (Advantage® Multi, Advocate®) for the Treatment of *Dirofilaria immitis* Circulating Microfilariae in Dogs. *Parasitol. Res.* **2015**, *114* Suppl. S1, S165–S174. [CrossRef] [PubMed]
16. Simparica Trio Zoetis Petcare. Available online: https://www.zoetispetcare.com/products/simparica-trio (accessed on 20 May 2021).
17. Snyder, P.W.; Hogg, J.T.; Ezenwa, V.O. Comparison of modified Flotac and Baermann techniques for quantifying lungworm larvae in free-ranging bighorn sheep (*Ovis canadensis*) feces, Montana, USA. *J. Wildl. Dis.* **2015**, *51*, 843–848. [CrossRef] [PubMed]
18. Schmahl, G.; Mehlhorn, H.; Harder, A.; Klimpel, S.; Krieger, K.J. Efficacy of a Combination of Imidacloprid plus Moxidectin against Larval and Adult Stages of Nematodes (*Trichuris muris*, *Angiostrongylus cantonensis*) in Rodents. *Parasitol. Res.* **2007**, *101* (Suppl. S1), 85–92. [CrossRef]
19. Hess, A.S.; Hess, J.R. Understanding tests of the association of categorical variables: The Pearson chi-square test and Fisher's exact test. *Transfusion* **2017**, *57*, 877–879. [CrossRef] [PubMed]
20. Neely, J.G.; Hartman, J.M.; Forsen, J.W.; Wallace, M.S.; Clinical Research Working Group. Tutorials in Clinical Research: Part VII. Understanding Comparative Statistics (Contrast)—Part A: General Concepts of Statistical Significance. *Laryngoscope* **2003**, *113*, 1534–1540. [CrossRef] [PubMed]

21. Bhaibulaya, M. Comparative studies on the life history of *Angiostrongylus mackerrasae* (Bhaibulaya, 1968) and *Angiostrongylus cantonensis* (Chen, 1935). *Int. J. Parasitol.* **1975**, *5*, 7–20. [CrossRef] [PubMed]
22. Buchter, V.; Hofmann, D.; Häberli, C.; Keiser, J. Characterization of Moxidectin against *Strongyloides ratti*: In Vitro and In Vivo Activity and Pharmacokinetics in the Rat Model. *ACS Infect. Dis.* **2020**, *7*, 1069–1076. [CrossRef] [PubMed]
23. Janko, C.; Geyer, J. Moxidectin has a lower neurotoxic potential but comparable brain penetration in P-glycoprotein-deficient CF-1 mice compared to ivermectin. *J. Veter. Pharmacol. Ther.* **2012**, *36*, 275–284. [CrossRef] [PubMed]
24. Nau, R.; Sörgel, F.; Eiffert, H. Penetration of Drugs through the Blood-Cerebrospinal Fluid/Blood-Brain Barrier for Treatment of Central Nervous System Infections. *Clin. Microbiol. Rev.* **2010**, *23*, 858–883. [CrossRef]
25. Damsch, S.; Eichenbaum, G.; Tonelli, A.; Lammens, L.; Bulck, K.V.D.; Feyen, B.; Vandenberghe, J.; Megens, A.; Knight, E.; Kelley, M. Gavage-Related Reflux in Rats: Identification, pathogenesis, and toxicological implications (review). *Toxicol. Pathol.* **2010**, *39*, 348–360. [CrossRef] [PubMed]
26. Helm, J.R.; Morgan, E.R.; Jackson, M.W.; Wotton, P.; Bell, R. Canine angiostrongylosis: An emerging disease in Europe. *J. Veter. Emerg. Crit. Care* **2010**, *20*, 98–109. [CrossRef] [PubMed]
27. Page, S.W. Chapter 10. Antiparasitic Drugs. In *Small Animal Clinical Pharmacology*; Maddison, J.E., Page, S.W., Church, D.B., Eds.; Saunders/Elsevier: Edinburgh, UK, 2008; pp. 210–214; ISBN 9780702028588.

pathogens

MDPI

Perspective

Management of Rat Lungworm Disease (Neuroangiostrongyliasis) Using Anthelmintics: Recent Updates and Recommendations

John Jacob, Argon Steel, Kathleen Howe and Susan Jarvi *

Department of Pharmaceutical Sciences, Daniel K. Inouye College of Pharmacy, University of Hawaii at Hilo, Hilo, HI 96720, USA

* Correspondence: jarvi@hawaii.edu; Tel.: +1-(808)-932-7701; Fax: +1-(808)-932-8117

Abstract: While there have been legitimate concerns in the past regarding the use of anthelmintics for the management of neuroangiostrongyliasis (rat lungworm disease), recent studies demonstrate that they can be considered safe and efficacious, particularly albendazole, which is regarded as the choice anthelmintic for its management. However, physician hesitancy to prescribe, as well as problems of availability persist, at least in Hawaii, which is considered the epicenter of this disease in the US. As a result, many patients suffer a diminished quality of life or even death. Here, we discuss recent studies that provide insights into new treatments and preventative interventions, which can be more rigorously used for the management of neuroangiostrongyliasis. In summary, results from recent studies suggest that albendazole and avermectins are beneficial for post-exposure management, pyrantel pamoate is beneficial as a post-exposure prophylactic, and levamisole is deserving of further study for the treatment of neuroangiostrongyliasis.

Keywords: rat lungworm; neuroangiostrongyliasis; treatment; anthelmintics; albendazole; pyrantel pamoate

check for updates

Citation: Jacob, J.; Steel, A.; Howe, K.; Jarvi, S. Management of Rat Lungworm Disease (Neuroangiostrongyliasis) Using Anthelmintics: Recent Updates and Recommendations. *Pathogens* **2023**, *12*, 23. https://doi.org/10.3390/pathogens12010023

Academic Editor: Marcello Otake Sato

Received: 1 December 2022
Revised: 20 December 2022
Accepted: 20 December 2022
Published: 23 December 2022

1. Introduction

The use of anthelmintics for the management of neuroangiostrongyliasis (rat lungworm disease) has been historically controversial due to the theoretical concern that killing *Angiostrongyliasis cantonensis* (rat lungworm) larvae that have migrated into the central nervous system (CNS) could trigger a severe inflammatory response, resulting in exacerbation of symptoms and further complications [1–3]. Such concerns have made many clinicians hesitant to consider such therapeutic interventions, despite the lack of actual evidence to validate this theory.

A further argument against the use of anthelmintics questions their necessity. Historically, the first human neuroangiostrongyliasis cases were reported from Asian countries and were associated with headaches and other mild symptoms, were typically self-resolving [4–6], or were successfully treated with only corticosteroids [4,7,8]. This led many clinicians in the West, particularly in the US, to expect a similar prognosis. In fact, this disease has been historically considered flu-like and self-resolving [4–6,9], and has given rise to the misperceptions among US clinicians that anthelminthic drugs were unnecessary or to wait until all other treatment options have been exhausted [10–12].

We believe that these arguments against the use of anthelmintics should be further examined. Primarily, it has become clear that cases in the US are often quite severe—to the extent of being fatal [10–13]. While the reasons for a geographical variation in disease severity remain unclear, the expectation that neuroangiostrongyliasis cases will self-resolve is no longer a tenable reason for delaying treatment. Secondly, it is worth noting that numerous other cerebral parasitic infections that can also cause eosinophilic meningitis are routinely treated with anthelmintics. For instance, a high dose of albendazole is used for the management of baylisascariasis, toxocariasis, gnathostomiasis, and cysticercosis (cestode), and similarly, a high dose of praziquantel is used for the management of paragonimiasis and schistosomiasis [14].

Finally, our understanding of anthelmintics used for neuroangiostrongyliasis has improved in recent years. Several in vitro, in vivo, and clinical studies have confirmed the efficacy of albendazole and pyrantel pamoate as treatments for neuroangiostrongyliasis, with avermectins and levamisole emerging as promising candidates [15–17]. These insights have provided us with a much-improved understanding of these drugs' safety and efficacy, allowing us to correct misconceptions and better manage this disease. Although the drugs mentioned above are classified as anthelmintics, each has unique therapeutic features and limitations, which we review below and summarize in Table 1.

1.1. *Albendazole*

Albendazole is a benzimidazole anthelmintic and, to date, is the most suitable anthelmintic for the management of neuroangiostrongyliasis due to its broad spectrum of nematocidal activity and ability to cross the blood–brain barrier (BBB) [18]. A systematic literature survey by Jacob et al. (2022a) [19] on the clinical outcomes associated with benzimidazole treatment found no evidence of albendazole resulting in cerebral inflammation or exacerbation of symptoms among patients with confirmed diagnoses of neuroangiostrongyliasis. This survey included an estimated 1034 patients and 2561 animals and provides highly supportive evidence for the safe and effective use of albendazole–corticosteroid co-therapy. The estimated dose of albendazole reported in these studies was approximately 15 mg/kg/day or 400 mg twice daily (with an average body weight of 60 kg) [19]. While bone marrow suppression and associated symptoms have been reported with long-term use of albendazole, overall, albendazole is generally considered to be a very safe anthelmintic [20]. Furthermore, no such side effects have been reported among neuroangiostrongyliasis patients [19].

Recent changes in clinical guidelines are also shifting in favor of using albendazole in the treatment of neuroangiostrongyliasis. Two hospitals in Australia (Sydney Children's Hospital and Children's Health Queensland Hospital) recommend using albendazole for the early management of neuroangiostrongyliasis in pediatrics [21,22]. Similarly, in the USA, Hilo Medical Center Hospital and the Hawaii Governor's Rat Lungworm Taskforce also endorse the use of albendazole for the management of neuroangiostrongyliasis in adults [23,24].

Other Issues with Albendazole

In addition to the therapeutic concerns discussed above, other pharmacoeconomic and availability issues continue to hinder the broader adoption of albendazole in treating neuroangiostrongyliasis. For example, albendazole is one of the most expensive drugs on the US market, with a price estimated between USD 200-250/unit dose [25]. Since the use of albendazole for neuroangiostrongyliasis is still controversial and has not been approved by the US FDA, most insurance companies will not cover the cost of albendazole. According to the guidelines mentioned above [23,24], treatment for neuroangiostrongyliasis requires albendazole to be administered two times a day (BID) for 2–3 weeks, which means that the patient will have to personally pay an amount between USD 6000-9000 (i.e., 24–36-unit doses) just to cover the cost of albendazole.

Additionally, even in the scenario where the clinician is willing to prescribe albendazole and the patient's insurance company is willing to cover its cost, due to its high price and relatively infrequent demand, most pharmacies do not stock sufficient quantities for adequate treatment. Many patients have experienced delayed access to albendazole due to the above reasons, resulting in life-long neurological sequelae and an associated decline in their quality of life (personal communications). Out of desperation, some of these patients have tried to self-medicate using veterinary formulations of albendazole, which are readily available and cost only a small fraction of the human formulations (USD

20-50). Given problems with self-diagnosis and calculating proper dosage, the use of veterinary anthelmintics in humans is highly problematic. Alternatively, some patients acquire personal stocks of albendazole by ordering it from countries such as India and Thailand, where the price of this drug is considerably lower.

We suggest that this availability issue could be resolved by having corporate pharmacies establish centralized stocking systems for albendazole in endemic areas, thus ensuring that sufficient quantities are routinely available.

1.2. Pyrantel Pamoate

Pyrantel pamoate's potential efficacy against *A. cantonensis* has warranted discussion in previous studies [15,17,26], and the in vivo efficacy of this drug has recently been evaluated in an experimental rat model [27]. The findings suggest pyrantel pamoate to be an effective post-exposure prophylactic against neuroangiostrongyliasis by reducing the worm burden as well as delaying the establishment of infection, thus providing time for the administration of albendazole [27]. However, it should be emphasized that pyrantel pamoate is a luminal drug with activity limited to the gastrointestinal tract (GIT) and is only efficacious while the parasite is within the GIT. Once the parasite has entered systemic circulation, the drug is clinically ineffective [28,29]. Upon release of the results of the in vitro study in 2021 [17], Hilo Medical Center Hospital, Hilo, Hawaii, USA, immediately adopted the use of pyrantel pamoate as a post-exposure prophylactic in their clinical treatment guidelines [24]. This guideline recommends administering pyrantel pamoate as instructed by the manufacturer (the same dosage as for pinworm management), which is typically 11 mg/kg, depending on the manufacturer [30].

Pyrantel pamoate is available over the counter (OTC) from most pharmacies with an estimated cost ranging between USD 10–20 per dose. Since the prophylactic activity of pyrantel pamoate against *A. cantonensis* is a recent discovery [27], clinical data are not yet available.

1.3. Ivermectin

According to the literature, ivermectin [31,32] and levamisole [33,34] are the most widely used anthelmintics after benzimidazoles for the management of neuroangiostrongyliasis, and both drugs appear efficacious.

Ivermectin does not directly kill the nematode; its paralyzing effect delays the progression of the infection and, to some extent, eradicates the parasite via immune responses and hepatic clearance. However, avermectins do not cross the BBB [35], and therefore, once *A. cantonensis* has entered the CNS, the drug is expected to be clinically ineffective. Thus, ivermectin is only efficacious during the early stages of infection when the parasite is within the GIT or systemic circulation. In theory, introducing ivermectin to an albendazole–corticosteroid co-therapy might produce a synergistic effect by paralyzing the nematode, slowing the progression of infection, and simultaneously increasing the exposure time to the nematocidal effects of albendazole [17]. Such synergistic effects of multiple anthelmintics have proven efficacious against Bancroftian filariasis, another parasitic nematode [36]. Future studies should investigate and compare the efficacy of the ivermectin–albendazole–corticosteroid cocktail with albendazole–corticosteroid co-therapy.

1.4. Levamisole

Experimental animal studies have shown levamisole to significantly reduce worm/larval burden, with the earliest administration (1–5 days post-infection) showing the most efficacy [37–41]. However, due to side effects such as agranulocytosis and its use as a cocaine adulterant, levamisole has been withdrawn from many global markets and is no longer available in many countries, including the US [42,43]. As shown in Table 1, since levamisole appears beneficial for the management of neuroangiostrongyliasis in humans [5,33], further research seems worthwhile.

Table 1. Summary of possible anthelmintic interventions for the management of rat lungworm disease (neuroangiostrongyliasis).

Anthelmintic	Mechanism of Action	Advantages	Disadvantages	References	Recommending Guidelines
Albendazole	Antimitotic	• Larva/wormicide: kills the parasite • Crosses the BBB	• Slow acting • Very expensive (in the USA) • May not be readily available for purchase • Prescription required	Jacob et al., 2021 [17] Jacob et al., 2022a [19] Jacob et al., 2022b [44]	Pediatric: • Children's Health Queensland Hospital [21] • Sydney Children's Hospital [22] Adult: • Hawaii Governor's Taskforce [23] • Hilo Medical Center Hospital [24]
Pyrantel pamoate	Nicotinic agonist	• Prophylactic • Rapid acting • Easily affordable • Prescription not required (over-the-counter)	• Works for only a short duration post-exposure * • Temporary paralysis of the parasite • Ineffective once the parasites are in the systemic circulation • Does not cross BBB	Jacob et al., 2021 [17] Jacob et al., 2022c [27]	• Hilo Medical Center Hospital [24]
Ivermectin	GABA agonist	• Rapid acting • Comparatively affordable	• May not be readily available for purchase • Prescription required • Temporary paralysis of the parasite • Does not cross BBB • Ineffective once the parasites are in the CNS	Jacob et al., 2021 [17] Monteiro et al., 2020 [31] Defo et al. , 2018 [32]	N/A
Levamisole	Nicotinic agonist	• Rapid acting	• Withdrawn from most global markets (Unavailable in the USA) • Prescription required • Temporary paralysis of the parasite • Does not cross BBB • Ineffective once the parasites are in the CNS	Jacob et al., 2021 [17] Hwang et al., 1994 [34] Ma et al., 2018 [33]	N/A

BBB: Blood–brain barrier; GABA: gamma-aminobutyric acid; CNS: Central nervous system; N/A: none available; * further research is required to estimate the time.

2. Conclusions

While there have been legitimate concerns in the past regarding the use of anthelmintics, recent studies demonstrate that they can be considered safe and efficacious for the management of neuroangiostrongyliasis. Additionally, these recent studies also provide insights into more effective management of neuroangiostrongyliasis. Furthermore, attention needs to be directed toward their pharmacoeconomic and availability aspects, which vary widely among these anthelmintics. In summary, results from past and current studies suggest that albendazole and avermectins are beneficial for post-exposure management, pyrantel pamoate is beneficial as a post-exposure prophylactic, and levamisole appears deserving of further research for the treatment of neuroangiostrongyliasis.

Funding: This work is supported by the Hawaii State Legislature and the Daniel K. Inouye College of Pharmacy.

Conflicts of Interest: The authors declare no conflict of interest.

References

1. Pien, F.D.; Pien, B.C. *Angiostrongylus cantonensis* Eosinophilic Meningitis. *Int. J. Infect. Dis.* **1999**, *3*, 161–163. [CrossRef] [PubMed]
2. Wang, Q.-P.; Lai, D.-H.; Zhu, X.-Q.; Chen, X.-G.; Lun, Z.-R. Human Angiostrongyliasis. *Lancet Infect. Dis.* **2008**, *8*, 621–630. [CrossRef]
3. Prociv, P.; Turner, M. Neuroangiostrongyliasis: The "Subarachnoid Phase" and Its Implications for Anthelminthic Therapy. *Am. J. Trop. Med. Hyg.* **2018**, *98*, 353–359. [CrossRef]
4. Khamsai, S.; Sawanyawisuth, K.; Senthong, V.; Limpawattana, P.; Chindaprasirt, J.; Intapan, P.M.; Maleewong, W.; Tiamkao, S.; Chotmongkol, V.; Ngamjarus, C. Corticosteroid Treatment Reduces Headache in Eosinophilic Meningitis: A Systematic Review. *Drug Target Insights* **2021**, *15*, 1–4. [CrossRef] [PubMed]
5. Hwang, K.P.; Chen, E.R. Clinical Studies on Angiostrongyliasis cantonensis among Children in Taiwan. *Southeast Asian J. Trop. Med. Public Health* **1991**, *22*, 194–199. [PubMed]
6. Diao, Z.; Wang, J.; Qi, H.; Li, X.; Zheng, X.; Yin, C. Treatment of Angiostrongyliasis Using a Combination of Albendazole and Dexamethasone: The Results of a Retrospective and Comparative Study. *Ann. Trop. Med. Parasitol.* **2011**, *105*, 65–69. [CrossRef]
7. Chotmongkol, V.; Sawanyawisuth, K.; Thavornpitak, Y. Corticosteroid Treatment of Eosinophilic Meningitis. *Clin. Infect. Dis.* **2000**, *31*, 660–662. [CrossRef]
8. Chotmongkol, V.; Sawanyawisuth, K. Clinical Manifestations and Outcome of Patients with Severe Eosinophilic Meningoencephalitis Presumably Caused by *Angiostrongylus cantonensis*. *Southeast. Asian J. Trop. Med. Public Health* **2002**, *33*, 231–234.
9. Saulo, A.A. *Avoid Contracting Angiostrongyliasis (Rat Lungworm Infection): Wash Fresh Fruits and Vegetables before Eating!* College of Tropical Agriculture and Human Resources: Honolulu, HI, USA, 2009.
10. Liu, E.W.; Schwartz, B.S.; Hysmith, N.D.; DeVincenzo, J.P.; Larson, D.T.; Maves, R.C.; Palazzi, D.L.; Meyer, C.; Custodio, H.T.; Braza, M.M.; et al. Rat Lungworm Infection Associated with Central Nervous System Disease—Eight U.S. States, January 2011–January 2017. *Morb. Mortal. Wkly. Rep.* **2018**, *67*, 825–828. [CrossRef]
11. Kliks, M.M.; Kroenke, K.; Hardman, J.M. Eosinophilic Radiculomyeloencephalitis: An Angiostrongyliasis Outbreak in American Samoa Related to Ingestion of Achatina fulica Snails. *Am. J. Trop. Med. Hyg.* **1982**, *31*, 1114–1122. [CrossRef]
12. Johnston, D.I.; Dixon, M.C.; Elm, J.L.; Calimlim, P.S.; Sciulli, R.H.; Park, S.Y. Review of Cases of Angiostrongyliasis in Hawaii, 2007–2017. *Am. J. Trop. Med. Hyg.* **2019**, *101*, 608–616. [CrossRef] [PubMed]
13. Martínez-Delgado, J.F.; González-Cortiñas, M.; Tápanes-Cruz, T.R.; Ruiz-Méndez, A. Meningoencefalitis Eosinofílica En Villa Clara (Cuba). Estudio de 17 Pacientes. *Rev. Neurol.* **2000**, *31*, 417–421. [CrossRef]
14. Graeff-Teixeira, C.; da Silva, A.C.A.; Yoshimura, K. Update on Eosinophilic Meningoencephalitis and Its Clinical Relevance. *Clin. Microbiol. Rev.* **2009**, *22*, 322–348. [CrossRef] [PubMed]
15. Mentz, B.M.; Graeff-Teixeira, C. Review Drug Trials for Treatment of Human Angiostrongyliasis. *Rev. Inst. Med. Trop. Sao Paulo* **2003**, *45*, 179–184. [CrossRef] [PubMed]
16. Roquini, D.B.; Silva, G.L.; Ferreira, L.L.G.; Andricopulo, A.D.; Wilairatana, P.; De Moraes, J. Susceptibility of *Angiostrongylus cantonensis* Larvae to Anthelmintic Drugs. *Front. Pharmacol.* **2022**, *13*, 1–9. [CrossRef]
17. Jacob, J.; Tan, G.; Lange, I.; Saeed, H.; Date, A.; Jarvi, S. In Vitro Efficacy of Anthelmintics on *Angiostrongylus cantonensis* L3 larvae. *Parasitology* **2021**, *148*, 240–250. [CrossRef]
18. Ramírez, T.; Benítez-Bribiesca, L.; Ostrosky-Wegman, P.; Herrera, L.A. In Vitro Effects of Albendazole and Its Metabolites on The Cell Proliferation Kinetics and Micronuclei Frequency of Stimulated Human Lymphocytes. *Arch. Med. Res.* **2001**, *32*, 119–122. [CrossRef]
19. Jacob, J.; Steel, A.; Lin, Z.; Berger, F.; Zöeller, K.; Jarvi, S. Clinical Efficacy and Safety of Albendazole and Other Benzimidazole Anthelmintics for Rat Lungworm Disease (Neuroangiostrongyliasis): A Systematic Analysis of Clinical Reports and Animal Studies. *Clin. Infect. Dis.* **2022**, *74*, 1293–1302. [CrossRef]

20. US Food and Drug Administration. *Prescribing Information Albenza*; US Food and Drug Administration: White Oak, MD, USA, 2009.
21. The Children's Health Queensland Hospital and Health Services Paediatric Guideline. Snail and Slug Ingestion. Available online: https://www.childrens.health.qld.gov.au/wp-content/uploads/PDF/ams/gdl-01219.pdf (accessed on 2 December 2022).
22. The Sydney Children's Hospital Network. Snail and Slug Ingestion Practice Guideline. Available online: https://www.schn.health.nsw.gov.au/_policies/pdf/2018-163.pdf (accessed on 2 December 2022).
23. Ansdell, V.; Kramer, K.J.; McMillan, J.K.; Gosnell, W.L.; Murphy, G.S.; Meyer, B.C.; Blalock, E.U.; Yates, J.; Lteif, L.; Smith, O.A.; et al. Guidelines for the Diagnosis and Treatment of Neuroangiostrongyliasis: Updated Recommendations. *Parasitology* **2021**, *148*, 227–233. [CrossRef]
24. Hilo Medical Center. Treatment Protocol for Rat Lungworm, Angiostrongyliasis. Available online: https://www.hilomedicalcenter.org/wp-content/uploads/2020/02/Rat-Lungworm-for-ED-Doctors-2020.pdf (accessed on 2 December 2022).
25. Shahriar, A.A.; Alpern, J.D. Antiparasitic Drugs in the United States—Two Roads to High Prices. *Front. Sociol.* **2020**, *5*, 540478. [CrossRef]
26. Lämmler, V.G.; Weidner, E. Zur Larviziden Wirkung von Anthelminthika Gegen *Angiostrongylus cantonensis. Berl. Munch. Tierarztl. Wochenschr.* **1975**, *88*, 152–156. [PubMed]
27. Jacob, J.; Steel, A.; Kaluna, L.; Hess, S.; Leinbach, I.; Antaky, C.; Sugihara, R.; Hamilton, L.; Follett, P.; Howe, K.; et al. In Vivo Efficacy of Pyrantel Pamoate as a Post-Exposure Prophylactic for Rat Lungworm (*Angiostrongylus cantonensis*). *Int. J. Parasitol. Drugs Drug Resist.* **2022**, *19*, 1–5. [CrossRef] [PubMed]
28. Martin, R.; Geary, T. Pharmacology of Pyrantel. In *Pyrantel Parasiticide Therapy in Humans and Domestic Animals*; Marchiondo, A., Ed.; Academic Press: Cambridge, MA, USA, 2016; pp. 21–45.
29. Pickering, L.; Baker, C.; Long, S.; McMillan, J. American Academy of Pediatrics. Pinworm Infection. In *Red Book: Report of the Committee on Infectious Diseases*; Pickering, L.K., Bakerm, C.J., Long, S.S., McMillan, J.A., Eds.; American Academy of Pediatrics: Itasca, IL, USA, 2006; pp. 520–522.
30. Papich, M.G. Pyrantel pamoate, pyrantel tartrate. In *Saunders Handbook of Veterinary Drug*, 4th ed.; Papich, M.G., Ed.; W.B. Saunders: St. Louis, MO, USA, 2016; pp. 693–694.
31. Monteiro, M.D.; Neto, E.G.D.C.; dos Santos, I.P.; Biernat, M.S.; Machado, R.M.; Hauser, V.B.; Rieder, C.R.; Spengler, C.N.; Fey, V.P.; Morassutti, A.; et al. Eosinophilic Meningitis Outbreak Related to Religious Practice. *Parasitol. Int.* **2020**, *78*, 102158. [CrossRef] [PubMed]
32. Defo, A.L.; Lachaume, N.; Cuadro-Alvarez, E.; Maniassom, C.; Martin, E.; Njuieyon, F.; Henaff, F.; Mrsic, Y.; Brunelin, A.; Epelboin, L.; et al. *Angiostrongylus cantonensis* Infection of Central Nervous System, Guiana Shield. *Emerg. Infect. Dis.* **2018**, *24*, 1153–1155. [CrossRef] [PubMed]
33. Ma, M.; Zhang, M.; Qiu, Z. Eosinophilic Meningitis Caused by *Angiostrongylus cantonensis* in An Infant: A case report. *Medicine* **2018**, *97*, e10975. [CrossRef]
34. Hwang, K.P.; Chen, E.R.; Chen, T.S. Eosinophilic Meningitis and Meningoencephalitis in Children. *Zhonghua Min. Guo Xiao Er Ke Yi Xue Hui Za Zhi* **1994**, *35*, 124–135.
35. Lovell, R.A. Ivermectin and Piperazine Toxicoses in Dogs and Cats. *Vet. Clin. North Am. Small Anim. Pract.* **1990**, *20*, 453–468. [CrossRef]
36. Ismail, M.; Jayakody, R.; Weil, G.; Nirmalan, N.; Jayasinghe, K.; Abeyewickrema, W.; Sheriff, M.; Rajaratnam, H.; Amarasekera, N.; De Silva, D.; et al. Efficacy of Single Dose Combinations of Albendazole, Ivermectin and Diethylcarbamazine for the Treatment of Bancroftian Filariasis. *Trans. R. Soc. Trop. Med. Hyg.* **1998**, *92*, 94–97. [CrossRef]
37. Hwang, K.P.; Chen, E.R. Anthelmintic Effect of Levamisole against *Angiostrongylus cantonensis* in mice. *Gaoxiong Yi Xue Ke Xue Za Zhi* **1994**, *10*, 536–542.
38. Kamath, V.R.; Menon, S.; Bhopale, M.K.; Deshpande, V.R.; Renapurkar, D.M. Experimental Chemotherapy of *Angiostrongylus cantonensis* Infection in Mice with Flubendazole, Levamisole, and Their Combination. *Polia Parasitol.* **1987**, *34*, 87–92.
39. Ishii, A.I.; Terada, M.; Fujiu, Y.; Sano, M. In Vivo Efficacy of Levamisole against Larval Stages of *Angiostrongylus cantonensis* and *A. costaricensis. Southeast Asian J. Trop. Med. Public Health* **1989**, *20*, 109–117. [PubMed]
40. Jindrak, K.; Alicata, J. Effect of 1-Tetramisole on *Angiostrongylus cantonensis* Infection in Rats. *Chemotherapy* **1969**, *14*, 244–252. [CrossRef] [PubMed]
41. Ishih, A. Die Wirkung Des Anthelminthikums Levamisol Auf ErsteAngiostrongylus Cantonensis-Larven in Infizierten Ratten. *Parasitol Res.* **1994**, *80*, 267–270. [CrossRef] [PubMed]
42. Chang, A.; Osterloh, J.; Thomas, J. Levamisole: A Dangerous New Cocaine Adulterant. *Clin. Pharmacol. Ther.* **2010**, *88*, 408–411. [CrossRef]
43. Solomon, N.; Hayes, J. Levamisole: A High Performance Cutting Agent. *Acad. Forensic Pathol.* **2017**, *7*, 469–476. [CrossRef]
44. Jacob, J.; Siraj, A.; Steel, A.; Tan, G.; Jarvi, S. Evaluation of the Mechanism of Action of Albendazole on Adult Rat Lungworm (*Angiostrongylus cantonensis*). *Exp. Parasitol.* **2022**, *242*, 108355. [CrossRef]

MDPI AG
Grosspeteranlage 5
4052 Basel
Switzerland
Tel.: +41 61 683 77 34

Pathogens Editorial Office
E-mail: pathogens@mdpi.com
www.mdpi.com/journal/pathogens

www.ingramcontent.com/pod-product-compliance
Lightning Source LLC
Chambersburg PA
CBHW041342210926
51850CB00013B/2957